KINETIC THEORY

Edited by **George Z. Kyzas**
and **Athanasios C. Mitropoulos**

Kinetic Theory
http://dx.doi.org/10.5772/intechopen.68734
Edited by George Z. Kyzas and Athanasios C. Mitropoulos

Contributors

Aiguo Xu, Guangcai Zhang, Yudong Zhang, Tatsuaki Tsuruyama, Nazish Rubab, Sadia Zaheer, Kashif Chaudhary, Viktor Gerasimenko, Leonid Fomin, Alexander Lokoshchenko, George Z. Kyzas, Athanasios Mitropoulos

Published by InTech

Janeza Trdine 9, 51000 Rijeka, Croatia

Publishing Process Manager Teo Kos
Technical Editor SPi Global
Cover InTech Design team

First published February, 2018
Printed in Croatia
Legal deposit, Croatia: National and University Library in Zagreb

Additional hard copies can be obtained from orders@intechopen.com

Kinetic Theory, Edited by George Z. Kyzas and Athanasios C. Mitropoulos
p. cm.
Print ISBN 978-953-51-3802-0
Online ISBN 978-953-51-3801-3

For EU product safety concerns:
IN TECH d.o.o., Prolaz Marije Krucifikse Kozulić 3, 51000 Rijeka, Croatia,
info@intechopen.com or visit our website at intechopen.com.

Meet the editor

Dr. George Z. Kyzas was born in Drama, Greece, and obtained his BSc degree in Chemistry and MSc and PhD degrees in Chemical Technology—Materials Science from the Aristotle University of Thessaloniki, Greece. His current interests include the synthesis of various adsorbent materials for the treatment of wastewaters (such as dyes, heavy metals, pharmaceuticals, and phenols). He has published over 85 significant scientific papers, authored books, edited books, and published chapters in books, teaching notes, and reports. He also acted as a guest editor in special issues of journals and presented many works in international conferences. He has been awarded with honors, grants, and fellowships for his research career/profile by the research committee of the Aristotle University of Thessaloniki, the National State Scholarships Foundation of Greece, and the Stavros Niarchos Foundation.

Prof. A. C. Mitropoulos was born in Athens in 1957. He received his BSc degree in Chemistry from the University of Thessaloniki and MSc and PhD degrees in Physical Chemistry from the University of Bristol. In 1998, he was appointed professor at the Department of Petroleum Engineering of the Eastern Macedonia and Thrace Institute of Technology. Since 2008, he has been the president of the same institute. His expertise is in the characterization of porous media, nanoporous materials, and membranes with in-situ techniques of adsorption and small angle X-ray scattering. He has published more than 100 journal papers, book chapters, and patents. He is a member of the Society of Petroleum Engineers.

Contents

Preface

This book "Kinetic Theory" describes many theories in kinetics. The world is governed by motions. The term kinetics partially originated from the Greek word "kinisis," which means motion. How important is motion in our life is easily understood. The kinetic theory describes a gas as a large number of submicroscopic particles (atoms or molecules), all of which are in constant rapid motion that has randomness arising from their many collisions with each other and with the walls of the container. Kinetic theory explains the macroscopic properties of gases, such as pressure, temperature, viscosity, thermal conductivity, and volume, by considering their molecular composition and motion. The theory posits that gas pressure is due to the impacts, on the walls of a container, of molecules or atoms moving at different velocities. Kinetic theory defines the temperature in its own way, in contrast to the thermodynamic definition. All the above clearly indicate that the "world" of kinetics has various sections. Therefore, the target of this book is wide. Its aim was the detailed analysis of kinetic theories in many scientific areas of recent research. Kinetic theory is not a unique term for gases, liquids, and so on. It is a complicated field gathering numerous topics as, for example, kinetic theory of gases, kinetic theory of liquids and vapors, thermodynamic aspects, transportation phenomena, adsorption-kinetic theories, linear and nonlinear kinetic equations, quantum kinetic theory, kinetic theory of nucleation, plasma kinetic theory, and relativistic kinetic theory. Specialists, researchers, and professors from more than 20 countries have published in this book their research into the kinetic theories/properties of various materials suitable for use in the optoelectronic devices, the development of new structures, and the results of their practical application. We are grateful to all the authors who have contributed their tremendous expertise to the present book, and we wish to acknowledge the outstanding support from Mr. Teo Kos, Publishing Process Manger, InTech Open Science Croatia, who collaborated tirelessly in crafting this book.

The future of kinetic theories is indeed bright!

Dr. George Z. Kyzas (MSc, PhD) and Prof. Athanasios C. Mitropoulos (MSc, PhD)
Hephaestus Advanced Laboratory
Eastern Macedonia and Thrace Institute of Technology
Kavala, Greece

Chapter 1

Introductory Chapter: Kinetics from Past to Future

George Z. Kyzas and Athanasios C. Mitropoulos

Additional information is available at the end of the chapter

http://dx.doi.org/10.5772/intechopen.73591

1. Introduction

The world is governed by motions. The term kinetics is partially originated from the Greek word "kinisis," which means motion. How important is motion in our life is easily understood. But how the kinetic theories have been developed during years? This question can be replied if some historic points are presented. So, to understand the development of kinetic theories, it is necessary to briefly give some history points during the analysis of kinetics.

2. The passage of kinetics during the decades

2.1. 19th Century

In 1850, Wilhelmy from Germany considered the rate of reversal of sucrose (hydrolysis into D-(+)- glucose and D-(−)- fructose within the sight of a corrosive) and observed it to be corresponding to the centralizations of both the sugar and the corrosive. Fourteen years later in 1864, two scientists from Norway (Guldberg and Waage) figured their "law of mass activity" as per which the response "powers" are corresponding to the result of the centralizations of the reactants: $K = [R]^r \cdot [S]^s/([A]^a\ [B]^b)$. In this equation, a, b, r, and s are some stoichiometric coefficients in the substance condition A + B = R + S. So the rate of the forward response is corresponding to $[A]^a\ [B]^b$ and that of the turnaround response is relative to $[R]^r\ [S]^s$.

In 1865, Harcourt and Esson (UK) examined the responses amongst H_2O_2 and Howdy and amongst $KMnO_4$ and $(COOH)_2$. They composed the relating differential conditions, coordinated them, and decided the fixation versus time connections. They additionally proposed a condition for the temperature reliance of the response rate, $k = A \cdot T^C$.

In 1884, van't Hoff (Netherlands) distributed his "Investigations of compound progression" (Études de dynamique chimique), in which he summed up and additionally built up crafted by Wilhelmy, Harcourt, and Esson. Specifically, he presented the differential technique for examination. He additionally dissected the temperature reliance of the balance steady (now called the "van't Hoff condition") and of forward and turnaround response rates.

In 1887, Ostwald (Germany; Latvia) presents the expressions "response request" and "half-life" in his "Lehrbuch der allgemeinen Chemie." Two years later, Arrhenius (Sweden) additionally broke down the temperature reliance of response rate, $k = A \cdot e^{-B/T}$, and gave it a "vitality obstruction" elucidation; this is presently called the "Arrhenius condition."

2.2. 20th Century

In the twentieth century, there have been huge advancements in the hypothesis of compound energy (assurance of rate constants and response orders from "first standards"). It is not yet conceivable, be that as it may, to foresee the motor parameters for true substance forms, and in reactor outline we should depend on precisely arranged and executed examinations. These hypothetical (and test) improvements are past the extent of a starting CRE course, with the exception of:

- 1913: Chapman (UK) presented and Bodenstein (Germany) built up the unfaltering state guess in chain responses, as indicated by which the rate of progress of halfway items is unimportant.
- 1917: Trautz (Germany) and Lewis (UK) autonomously suggested that the rate of response is controlled by the recurrence of atomic impacts. This is currently known as the "impact hypothesis" of substance response energy.
- 1920s: Langmuir (USA) contemplated the energy of surface responses and presented what is currently known as the "Langmuir isotherm" which was additionally created by Hinshelwood (UK) into the "Langmuir-Hinshelwood system" of heterogeneous responses.
- 1934: Rice and Herzfeld (USA) demonstrate that chain responses including free radicals (whose fixations are resolved utilizing the consistent state estimate) are in charge of the regularly watched varieties in the request ($n = 0.5$, $n = 1$, $n = 1.5$, and so forth) of warm disintegration of natural mixes, for example, ethane and acetaldehyde.
- 1935: Eyring (USA) built up a measurable treatment called the "hypothesis of supreme response rates" or "change state hypothesis," as per which the response happens in two stages: (an) equilibrated transformation of the reactant(s) into an "initiated complex"; (b) decay of the mind boggling (which happens at an unmistakable rate).

3. Conclusions

All above clearly indicate that the "world" of kinetics has various sections. Therefore, the target of this Book should be wide. Its aim was the detailed analysis of kinetic theories in many scientific areas of recent research. Kinetic theory is not a unique term for gases, liquids etc. It is a complicated field gathering numerous topics as for example: kinetic theory of gases; kinetic theory of liquids, vapors; thermodynamic aspects; transportation phenomena; adsorption-kinetic theories; linear and nonlinear kinetic equations; quantum kinetic theory; kinetic theory of nucleation; plasma kinetic theory; and relativistic kinetic theory.

Author details

George Z. Kyzas* and Athanasios C. Mitropoulos

*Address all correspondence to: georgekyzas@gmail.com

Hephaestus Advanced Laboratory, Eastern Macedonia and Thrace Institute of Technology, Kavala, Greece

Chapter 2

Discrete Boltzmann Modeling of Compressible Flows

Aiguo Xu, Guangcai Zhang and Yudong Zhang

Additional information is available at the end of the chapter

http://dx.doi.org/10.5772/intechopen.70748

Abstract

Mathematically, the typical difference of discrete Boltzmann model (DBM) from the traditional hydrodynamic one is that the Navier-Stokes (NS) equations are replaced by a discrete Boltzmann equation. But physically, this replacement has a significant gain: a DBM is roughly equivalent to a hydrodynamic model supplemented by a coarse-grained model of the thermodynamic non-equilibrium (TNE) effects, where the hydrodynamic model can be and can also be beyond the NS. Via the DBM, it is convenient to perform simulations on systems with flexible Knudsen number. The observations on TNE are being obtaining more applications with time.

Keywords: Boltzmann equation, discrete Boltzmann modeling, Navier-Stokes, compressible flows, non-equilibrium effects

1. Introduction

Generally, compressible flow is frequently referred also to as gas dynamics which is the branch of fluid mechanics that deals with flows having significant changes in fluid density. That is because gases, mostly, display such behavior. In fact, all materials are compressible. Besides liquids, gases and plasmas, to some extent, the plastic solids under strong shock can also be modeled as compressible flows. Because, in the last case, that the strength of material is negligibly smaller than that of the shock. Flows with a Mach number less than 0.3 are usually treated as being incompressible for that the variation of density due to velocity is less than 5% in that case. The study on compressible flow is relevant to various fields, such as high-speed aircraft, rocket motors, jet engines, high-speed entry into a planetary atmosphere, gas pipelines, etc. where shock wave and/or detonation play a significant role. In this chapter, the following several kinds of flows including high Mach number flows with combustion, multiphase flows with phase separation and complex flows with hydrodynamic instability[1]

[1]It is clear that there exist significant overlaps among those classifications.

are taken as examples, and all flows are treated uniformly as being compressible. It is hopeful that the methods and ideas developed in this chapter may be adapted, more or less, to other kinds of complex flows.

Some common and typical features of these flows are as below: each of them possesses multi-scale structures and/or kinetic modes. There exist plenty of interfaces inside the system. The interfaces include material interfaces and mechanical interfaces. Each of them experiences complicated competitions between behaviors in various spatial-temporal scales. The forcing and responsive processes inside the system are very complicated. Such a system generally shows pronounced non-equilibrium behaviors.

1.1. Traditional models

The traditional macroscopic models of compressible flows are generally based on Navier-Stokes (NS) equations, even Euler equations. The model of Euler equations assumes that the system is always at its local thermodynamic equilibrium (LTE). The NS model considers the thermodynamic non-equilibrium (TNE) via the viscous stress and heat flux. The viscous stress and heat flux compose a set of convenient and effective description of the TNE. But the description is also quite dense and coarse-grained. Many specific TNE behaviors are invisible under NS description, even though they are helpful for understanding the specific TNE status. Besides, since it includes only the first-order term of Knudsen number in the Chapman-Enskog expansion [1], it is reasonable only when the Knudsen number is very small. It cannot be used to access deeper TNE behaviors. To access the complicated non-equilibrium behaviors, one possible solution is to use the molecular dynamics (MD) [2] or direct simulation Monte Carlo (DSMC) [3]. The MD simulation can help to understand some fundamental mechanisms from the atomic level. But the spatial and temporal scales it can access are too small to be comparable with experiments. The DSMC simulation has a similar constraint of computational cost.

The NS model is not enough to capture so complex non-equilibrium behaviors while the MD and DSMC cannot access spatial-temporal scales that are large enough. Under such conditions, a kinetic approach based on the non-equilibrium statistical mechanics (NESM) [1, 4, 5] is preferable.

1.2. Non-equilibrium statistical mechanics

Statistical mechanics is a branch of theoretical physics. It was developed to study the average behavior of a mechanical system with uncertain state by using probability theory. Microscopic mechanical laws do not contain concepts such as temperature, heat, or entropy; however, statistical mechanics shows how these concepts arise from the natural uncertainty about the state of a system when that system is prepared in practice. Statistical mechanics provides exact methods to connect thermodynamic quantities to microscopic behavior.

The NESM is based on mechanics and some necessary assumptions. The concept of macroscopic observation and assumption of coarse-grained density are cornerstones. The Liouville equation [4, 5] is the most fundamental governing equation when without quantum fluctuations. It describes the N-particle system using a partial-differential equation for the probability density function, $f=f(\xi_1, \xi_2, \cdots, \xi_N, t)$, in a $6N$-dimensional space,

$$\frac{\partial f}{\partial t}+\sum_{i=1}^{N}\left(\frac{\boldsymbol{p}_i}{m}\cdot\frac{\partial}{\partial \boldsymbol{q}_i}-\frac{\partial}{\partial \boldsymbol{q}_i}\Phi^N\cdot\frac{\partial}{\partial \boldsymbol{p}_i}\right)f=0, \tag{1}$$

where m is the particle mass, $\zeta_i=(\boldsymbol{q}_i, \boldsymbol{p}_i)$, $\boldsymbol{q}_i$ and $\boldsymbol{p}_i$ are the coordinate and momentum of the ith particle, $\Phi^N(\boldsymbol{q}_1, \ldots, \boldsymbol{q}_N)$ is the interaction potential of all the N particles. By integration over part of the variables, the Liouville equation can be transformed into a chain of equations, which is referred to as the Bogoliubov-Born-Green-Kirkwood-Yvon (BBGKY) hierarchy [4], where the first equation connects the evolution of one-particle probability density function with the two-particle probability density function, the second equation connects the two-particle probability density function with the three-particle probability density function, and generally the Sth equation connects the S-particle probability density function, f_s,with the $(S+1)$-particle probability density function, f_{s+1}. Specifically,

$$\frac{\partial f_S}{\partial t}+\sum_{i=1}^{S}\left(\frac{\boldsymbol{p}_i}{m}\cdot\frac{\partial}{\partial \boldsymbol{q}_i}-\frac{\partial}{\partial \boldsymbol{q}_i}\Phi(\boldsymbol{q}_1,\ldots,\boldsymbol{q}_s)\cdot\frac{\partial}{\partial \boldsymbol{p}_i}\right)f_S=\frac{N-S}{V}\int\sum_{i=1}^{S}\left(\nabla_{\mathbf{r}_i}\Phi_{i,S+1}\right)\cdot\left(\nabla_{\boldsymbol{p}_i}f_{S+1}\right)d^6\zeta_{S+1}, \tag{2}$$

where V is the volume of system, $\Phi(\boldsymbol{q}_1,\ldots,\boldsymbol{q}_s)=\sum_{i<j\le s}\Phi_{i,j}\left(\boldsymbol{q}_i,\boldsymbol{q}_j\right)$ is the total two-particle interaction potential, $\Phi_{i,S+1}$ is the interaction potential between the particle i and particle $S+1$.

Here

$$V^n=\int_\Gamma f_n(\zeta_1,\zeta_2,\cdots,\zeta_n,t)d^6\zeta_1\cdots d^6\zeta_n, \tag{3}$$

and

$$f_S(\zeta_1,\zeta_2,\cdots,\zeta_S,t)=V^{(S-N)}\int f_N(\zeta_1,\zeta_2,\cdots,\zeta_N,t)d^6\zeta_{S+1}\cdots d^6\zeta_N. \tag{4}$$

When $S=N$, the above equation recovers to the Liouville equation. For a macroscopic system, the particle number, N, is in the order of Avogadro constant, 10^{23}. Generally, there is no way to solve Eq. (1) or Eq. (2).

We need to simplify the model via considering some simpler cases. If considering only the case where correlations among three and more particles are negligible, the two-particle interaction is relevant to their distance, and the two-particle distribution function can be written as the product of two single particle distribution functions, specifically,

$$\Phi_{1,2}(\boldsymbol{q}_1,\boldsymbol{q}_2)=\Phi_{1,2}(|\boldsymbol{q}_1-\boldsymbol{q}_2|), \tag{5}$$

$$f_2(\boldsymbol{q}_1,\boldsymbol{p}_1;\boldsymbol{q}_2,\boldsymbol{p}_2;t)=f_1(\boldsymbol{q}_1,\boldsymbol{p}_1;t)f_1(\boldsymbol{q}_2,\boldsymbol{p}_2;t), \tag{6}$$

we obtain the Boltzmann equation [1, 4],

$$\frac{\partial f}{\partial t}+\boldsymbol{v}\cdot\frac{\partial f}{\partial \boldsymbol{r}}+\boldsymbol{a}\cdot\frac{\partial f}{\partial \boldsymbol{v}}=Q(f,f), \tag{7}$$

with

$$Q(f,f) = \int_{-\infty}^{+\infty}\int_{0}^{4\pi} (f^* f_1^* - f f_1)\, g\, \sigma\, d\Omega\, dv_1. \tag{8}$$

Here is the single particle distribution function. Compared with the MD and Liouville equation, the Boltzmann equation is a much coarse-grained model. The purposes of establishing Boltzmann equation are to define and calculate the entropy in non-equilibrium state, to derive, prove, even modify the fundamental hydrodynamic differential equation [5].

1.3. Non-equilibrium statistical mechanics and macroscopic description

All hydrodynamic quantities are some kinds of kinetic moments of the distribution function and can be expressed as:

$$\mathbf{W} = \int f \begin{bmatrix} 1 \\ v \\ v \cdot v/2 \end{bmatrix} dv, \tag{9}$$

where the particle mass has been assumed to be 1, $\mathbf{W}$ is the vector of macroscopic quantity composed of density, momentum and energy, $\mathbf{W} = [\rho, \rho\mathbf{u}, \rho E]^T$. Besides, the total stress $\boldsymbol{\sigma}$, viscous stress $\boldsymbol{\Pi}$ and heat flux $\mathbf{q}$ are related to f via

$$\boldsymbol{\sigma} = \int f\,(v - \boldsymbol{u})(v - \boldsymbol{u})dv = \boldsymbol{\Pi} + p\boldsymbol{I}, \tag{10}$$

$$\boldsymbol{\Pi} = \int (f - f^{eq})\,(v - \boldsymbol{u})(v - \boldsymbol{u})dv, \tag{11}$$

and

$$\boldsymbol{q} = \frac{1}{2}\int (f - f^{eq})(v - \boldsymbol{u})(v - \boldsymbol{u}) \cdot (v - \boldsymbol{u})dv, \tag{12}$$

respectively. Compared with Boltzmann equation, in Navier-Stokes equations,

$$\frac{\partial \rho}{\partial t} + \nabla \cdot (\rho \boldsymbol{u}) = 0, \tag{13}$$

$$\frac{\partial (\rho \boldsymbol{u})}{\partial t} + \nabla \cdot (p\boldsymbol{I} + \rho \boldsymbol{u}\boldsymbol{u}) = -\nabla \cdot \boldsymbol{\Pi}, \tag{14}$$

$$\frac{\partial}{\partial t}(\rho E) + \nabla \cdot [(\rho E + p)\boldsymbol{u})] = -\nabla \cdot [\boldsymbol{q} + \boldsymbol{\Pi} \cdot \boldsymbol{u}], \tag{15}$$

where $\boldsymbol{\Pi} = 2/3\mu(\nabla \cdot \boldsymbol{u})\boldsymbol{I} - \mu(\nabla \boldsymbol{u})^T - \mu\nabla \boldsymbol{u}$, $\boldsymbol{q} = -\kappa\nabla T$, T is the temperature, μ and κ are the coefficients of viscosity and heat conductivity, respectively. More details of molecular motions are neglected and the distribution function f disappears as well. What remained are the conserved kinetic moments, density ρ, momentum (density) $\rho\boldsymbol{u}$ and energy(density) ρE, and a few non-conserved quantities, momentum fluxes $\rho\boldsymbol{uu}$ and $\boldsymbol{\Pi}$, energy fluxes $\rho E\boldsymbol{u}$ and $\boldsymbol{q}$. The relation between internal energy ρT and pressure p is given by the equation of state. In Euler equations, the viscous stress $\boldsymbol{\Pi}$ and heat flux $\boldsymbol{q}$ are further neglected.

From molecular dynamics to Boltzmann equation, to Navier-Stokes, and further to Euler equations, in each step, the description becomes coarse-grained, and the contained physical information becomes less [6]. The switching of model in each step corresponds to the state of system under consideration gets closer to its thermodynamic equilibrium, the behavior is simpler, consequently the system can be described by fewer physical variables. For a given non-equilibrium system, the switching of model in each step corresponds to the spatial-temporal scale that we use to observe the system becomes larger, consequently more smaller structures and quicker kinetic modes are invisible. What obtained are the remaining larger structures and slower kinetic modes. Based on Boltzmann equation, the most relevant TNE effects accompanying the hydrodynamic behaviors can be studied, in addition to the general hydrodynamic behaviors described by the hydrodynamic model.

2. Discrete Boltzmann theory

2.1. Discrete Boltzmann modeling

From Boltzmann equation to DBM, two steps of coarse-grained physical modelings are needed. The principle for coarse-grained modeling is that the physical quantities used to measure the system must keep the same values after simplification.

Step 1: Linearization of the collision term

Even though, compared with MD or Liouville equation, Boltzmann equation is a much coarse-grained model, its collision term is still too complicated to be solved in most practical cases. The simplest way to simplify is to introduce a local equilibrium velocity distribution function, f^{eq}, and write the collision term into the following linear form,

$$\frac{\partial f}{\partial t} + \boldsymbol{v} \cdot \frac{\partial f}{\partial \boldsymbol{r}} + \boldsymbol{a} \cdot \frac{\partial f}{\partial \boldsymbol{v}} = -\frac{1}{\tau}(f - f^{eq}). \tag{16}$$

The physical meaning of the linearized collision model is thus: collisions of molecules result in that f approaches to thermodynamic equilibrium f^{eq} and the relaxation time is controlled by the parameter τ. If we do not aim to measure the system using the specific values of f, but using only some of its kinetic moments, then only if these kinetic moments keep invariable after the simplification, that will be OK. The linearization of the collision term requires

$$\int -\frac{1}{\tau}(f - f^{eq}) \boldsymbol{\Psi} d\boldsymbol{v} = \int Q(f, f) \boldsymbol{\Psi} d\boldsymbol{v}, \tag{17}$$

where $\boldsymbol{\Psi} = [1, \boldsymbol{v}, \boldsymbol{vv}, \boldsymbol{vvv}, \cdots]^T$ contributes the kinetic moments used to measure the system. The specific form of f^{eq} depends on the terms that $\boldsymbol{\Psi}$ takes. According to the specific form of $\boldsymbol{\Psi}$ or f^{eq}, the linearized collision model may be referred to as the Bhatnagar-Gross-Krook (BGK) model [7–10], ellipsoidal statistical BGK model [11], Shakhov model (for monatomic gas) [12], Rykov model (for diatomic gas) [13], Liu model [14], etc. When only the mass, momentum and energy conservation laws are kept, f^{eq} takes the simplest form, the Maxwellian. This is the so-called BGK model. Due to its simplicity, the BGK model is most extensively used.

It should be note that the Single-Relaxation-Time model works when all the kinetic modes approaching to thermodynamic equilibrium share more or less the same relaxation time. For more complicated cases where the relaxation times of different kinetic modes approaching to thermodynamic equilibrium are significantly different, the multiple-relaxation-time (MRT) collision model is needed [15].

Step 2. Discretization of the particle velocity space

We first consider the case without the force term. The discrete Boltzmann equation reads,

$$\frac{\partial f_i}{\partial t} + v_{i\alpha}\frac{\partial f_i}{\partial r_\alpha} = -\frac{1}{\tau}\left(f_i - f_i^{eq}\right), \tag{18}$$

where i is the index of discrete velocity. Since the common schemes for discretizing the space and time do not work for discretizing the particle velocity space. To find an effective means to discretize the particle velocity space, we go back to consider what we really need. Here, we do not aim to describe the system using specific values of the discrete distribution function f_i, but its kinetic moments. So, only if these kinetic moments, originally in integral form, can be equally calculated in summation form, that will be acceptable. Specifically,

$$\int f\,\boldsymbol{\Psi}'(\boldsymbol{v})d\boldsymbol{v} = \sum_i f_i\,\boldsymbol{\Psi}'(\boldsymbol{v}_i), \tag{19}$$

where the left hand side gives the kinetic moments of f needed to describe the system and $\boldsymbol{v}_i$ at the right hand side is the discrete particle velocity. According to the Chapman-Enskog analysis, a kinetic moment of f can be finally calculated via an appropriate kinetic moment of f^{eq}. Therefore the requirement of Eq. (18) can be further written as

$$\int f^{eq}\,\boldsymbol{\Psi}''(\boldsymbol{v})d\boldsymbol{v} = \sum_i f_i^{eq}\boldsymbol{\Psi}''(\boldsymbol{v}_i), \tag{20}$$

where the left hand side gives the kinetic moments of f^{eq} being involved in the process of constructing DBM. The requirement of Eq. (20) can be rewritten as a matrix equation,

$$\hat{\boldsymbol{f}}^{eq} = \boldsymbol{M}\cdot \boldsymbol{f}^{eq}, \tag{21}$$

where $\hat{\boldsymbol{f}}^{eq} \equiv \left[\hat{f}_k^{eq}\right]$, $\boldsymbol{f}^{eq} \equiv [f_i^{eq}]$, $\boldsymbol{M} \equiv [m_{ki}]$, $\hat{f}_k^{eq}$ is the kth kinetic moment of $\boldsymbol{f}^{eq}$. The elements, m_{ki}, are determined by the discrete velocities. That is to say, the discrete velocities should be chosen in such a way that the requirement of Eq. (20) is satisfied. Under such a constraint, the discretization of particle velocity space is flexible. In fact, $\boldsymbol{M}$ can also be non-squared rectangular matrix. The choice of discretization scheme depends on the compromise between the numerical cost, stability and physical gain (such as physical symmetry).

Via the same idea, it is straight forward to formulate MRT-DBM. To ensure the relaxation times have clear physical meanings, in the MRT-DBM, the collision term is first calculated in the kinetic moment space, and then transformed back to the discrete velocity space. To ensure DBM describe reasonable flow behaviors, a correction term is needed [15].

According to the Chapman-Enskog analysis, to access system which is deeper into thermodynamic non-equilibrium, higher order terms in Knudsen number should be considered. As a result, the requirement of f in Eq. (19) will lead to more kinetic moment relations of f^{eq} in Eq. (20). Consequently, more discrete velocities are needed. For the case with inter-particle force, one should generally first finish the derivative calculation of f with respect to v, and then perform the discretization of particle velocity space.

2.2. Non-equilibrium: definition and measuring

Once the concept of equilibrium is defined, the concept of non-equilibrium is clear. In classical mechanics, a particle is in mechanical equilibrium if the net force on that particle is zero. By extension, a physical system made up of many parts is in mechanical equilibrium if the net force on each of its individual parts is zero. In addition to defining mechanical equilibrium in terms of force, there are many alternative definitions for mechanical equilibrium which are all mathematically equivalent. In terms of momentum, a system is in equilibrium if the momentum of its parts is all constant. In terms of velocity, the system is in equilibrium if velocity is constant. In a rotational mechanical equilibrium, the angular momentum of the object is conserved and the net torque is zero. More generally in conservative systems, equilibrium is established at a point in configuration space where the gradient of the potential energy with respect to the generalized coordinates is zero. Similarly, a fluid system is in fluid mechanical equilibrium or hydrodynamic equilibrium if the net force on each of its 'fluid particles' (small fluid elements) is zero and without temperature gradient around the 'fluid particle'.

For ideal gas system, in thermodynamic equilibrium there are no net macroscopic flows of matter or energy either within a system or between systems. In non-equilibrium systems, by contrast, there are net flows of matter or energy. Global thermodynamic equilibrium means that the relevant intensive parameters are homogeneous throughout the whole system, while local thermodynamic equilibrium means that those intensive parameters are varying in space and time, but are varying so slowly that, for any point, one can assume thermodynamic equilibrium in some neighborhood about that point. Rarefied gases at ordinary temperatures behave very nearly like ideal gas and the Maxwell speed distribution is an excellent approximation for such gases. Thus, it forms the basis of the kinetic theory of gases.

It is evident that Euler equations are used to investigate fluid flows which are at local thermodynamic equilibrium but mechanical non-equilibrium, while NS, Burnett and Super-Burnett equations are used to investigate fluid flows which are at mechanical and thermodynamic non-equilibrium. Only when all kinds of kinetic moments of $(f - f^{eq})$ are zero, the system is at thermodynamic equilibrium. hydrodynamic non-equilibrium (HNE) drives the evolution of f and results in viscous stress and heat flux.

In many practical cases, it is neither necessary to know all the details of f nor necessary to know all the kinetic moments of $(f-f^{eq})$.Therefore, one can (i) care only the main feature of f, specifically, neglect the high order terms of Knudsen number, (ii) care only the kinetic moments of $(f-f^{eq})$ which are most relevant to the macroscopic behaviors under consideration, specifically, those involved in constructing the discrete Boltzmann model.

A centrally important motivation of DBM is to check, measure and analyze the non-equilibrium state and effects [6, 16–18]. The DBM presents two sets of measures for the TNE. One set is dynamically from the difference of the kinetic moments of f and f^{eq},

$$\boldsymbol{\Delta}_m = \boldsymbol{M}_m(f) - \boldsymbol{M}_m(f^{eq}), \tag{22}$$

where $\boldsymbol{M}_m(f)$ is the mth rank moment of velocity tensor of f, $\boldsymbol{\Delta}_m^*$ can be obtained when $\boldsymbol{M}_m(f)$ is the mth rank central moment. The other set includes the viscous stress and heat flux. The former describes the specific TNE status, while the latter describes the influence of those TNE to the macroscopic control equations. The former is local, while the latter is non-local. The former is finer, while the latter is coarser. At each step of DBM simulation, both the f and f^{eq} are calculated. Therefore, the TNE effects are naturally included in each step. The captured TNE effects are just those being most relevant to the macroscopic flow behaviors under consideration.

Entropy production is a highly concerned quantity in both physics and engineering studies. From the physics side, it is helpful for understanding the complex non-equilibrium behaviors. From the engineering side, a process with lower entropy production may have higher energy transformation efficiency. Following the way of defining entropy equilibrium equation in the non-equilibrium thermodynamics, a new entropy equilibrium equation can be obtained as follows [19],

$$\frac{\partial s}{\partial t} = -\nabla \cdot \boldsymbol{J}_s + \sigma, \tag{23}$$

where s is the entropy density,

$$\boldsymbol{J}_s = s\boldsymbol{u} + \frac{1}{T}\boldsymbol{\Delta}_{3,1}^* \tag{24}$$

and

$$\sigma = \boldsymbol{\Delta}_{3,1}^* \cdot \nabla\frac{1}{T} - \frac{1}{T}\boldsymbol{\Delta}_2^* : \nabla\boldsymbol{u} + \rho\frac{Q}{T}F(\lambda) \tag{25}$$

are the entropy flux and entropy production rate, respectively. $F(\lambda) = \dot{\lambda}$ is a rate function describing the process of chemical reaction occurring in the system, where λ is defined as the ratio of the product density to the overall density. From Eq. (24) one can find that the entropy production results from three kinds of resources, the non-organized energy flux (NOEF), non-organized momentum flux (NOMF), and chemical reaction. From the relation, one can study the various mechanisms resulting in increase of entropy and their relative importance [19].

The TNE behaviors are very complex and difficult to quantitatively investigate. Finding a convenient and efficient method to characterize the TNE status and effects is the corner stone, DBM presents such a potential approach [16–25].

2.3. DBM versus CFD

The traditional CFD needs to first know the exact form of the hydrodynamic equations, then design numerical algorithm according to the properties of those equations. The DBM is a coarse-grained model derived from the Boltzmann equation. In principle, it can be formulated and applied to simulate flows without knowing the exact form of the hydrodynamic equations, only if necessary kinetic moments of f^{eq} are followed. From the perspective of physical application, a DBM is roughly equivalent to a hydrodynamic model supplemented by a coarse-grained model of TNE. Via the DBM, it is easy to perform multi-scale simulations in a wide range of Knudsen number.

3. Applications

3.1. Combustion system

Combustion has long been playing a dominant role in the transportation and power generation. To improve combustion efficiency and decrease pollution, in recent years, some new combustion concepts have been proposed. For example, pulsed detonation engine, spinning detonation engine, microscale combustion, nanopropellent, partially premixed and stratified combustion, plasma-assistant combustion, cool flames, etc. All these new combustion concepts involve complicated non-equilibrium chemical and transport processes [26].

The chemical reaction process is very complex and may include varieties of reaction mechanism. So far, most of the chemical reaction kinetic models are phenomenological. As the first step, we consider only the simplified form of Lee-Tarver chemical reaction rate law [21]. Considering the thermal initiation, the reaction kinetic is described by

$$\frac{d\lambda}{dt} = \begin{cases} a(1-\lambda) + b(1-\lambda)\lambda, & T \geq T_{th} \text{ and } 0 \leq \lambda \leq 1 \\ 0, & \text{else,} \end{cases} \tag{26}$$

where a and b are constants, λ is the concentration of the product and works as the reaction process parameter, T_{th} is the temperature threshold for chemical reaction. Consider the case where the chemical reaction is slow enough compared with the kinetic process of approaching thermodynamic equilibrium, so we can treat f as f^{*eq} during the reaction process. The evolution equation of single-relaxation-time DBM for combustion reads

$$\frac{\partial f_i}{\partial t} + v_{i\alpha}\frac{\partial f_i}{\partial r_\alpha} = -\frac{1}{\tau}\left(f_i - f_i^{*eq}\right), \tag{27}$$

where $f_i^{*eq} = f_i^{eq}(\rho, u, T + (\gamma - 1)\tau Q F(\lambda))$ is the local equilibrium distribution function taking into account chemical reaction effect. Q is the amount of heat released by the chemical reactant per unit mass. If the relaxation times of different kinetic modes approaching to thermodynamic equilibrium are significantly different, one needs the MRT model [22]. Some observations brought by DBM are as below.

Non-equilibrium quantities defined in Eq. (22) are used to study a simple case of detonation [21]. For the case of CJ detonation shown in **Figure 1(a)**, the corresponding non-equilibrium quantities, Δ_2^* and Δ_3^*, are shown in **Figure 1(b)** and **(c)**, respectively. Interestingly, at the position of von Neumann pressure peak, the system is not far from but is nearly in its thermodynamic equilibrium. The internal energies in two degrees of freedom deviate from their mean value in opposite direction with the same amplitude. The internal energy in each degree of freedom deviates from the mean value in opposite direction before and after the von Neumann pressure peak. The amplitude in front of the von Neumann pressure peak is larger. The physical reasons are as follows. The mechanical non-equilibrium is the driving force for TNE. When a strong shock comes, the density, temperature and flow velocity increases abruptly so that the system does not have enough time to relax to its thermodynamic equilibrium. With decrease of the gradients of density, temperature and flow velocity, the system gets relatively more time for thermodynamic relaxation. At the position of von Neumann pressure peak, the system has been close to its thermodynamic equilibrium. After the von Neumann pressure peak, the density and flow velocity decrease quickly so that the system does not have enough time for thermodynamic relaxation again. The total deriving force for TNE in front of the von Neumann pressure peak is larger than behind.

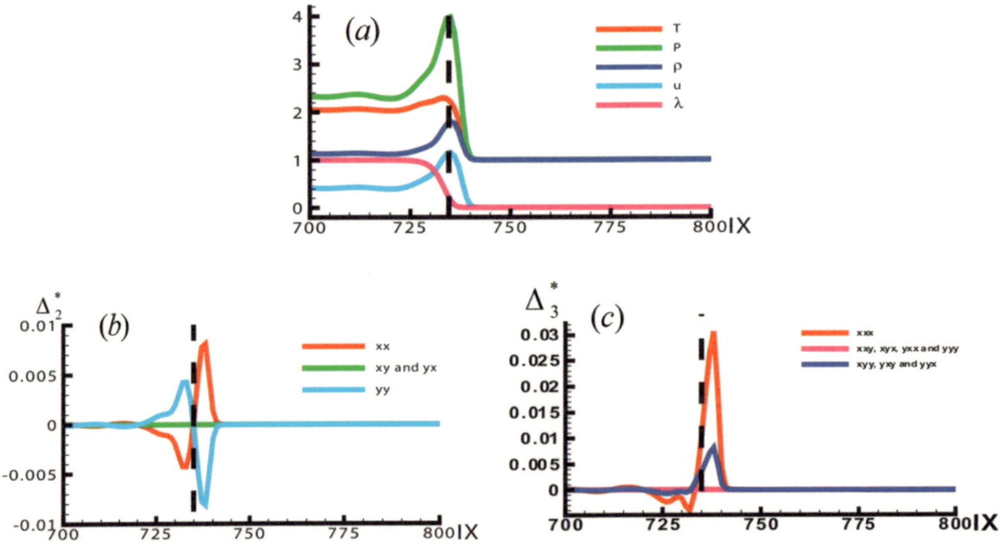

Figure 1. Profiles of physical quantities for CJ detonation. (a) The density ρ, pressure P, temperature T, x-component of velocity u, and the reaction process λ. (b) and (c) The non-equilibrium quantities Δ_2^* and Δ_3^*, respectively (see Ref. [21] for more details).

Figure 2 gives the pressure P and corresponding non-equilibrium quantity $\Delta_{2,xx}$ at a certain time [22]. From **Figure 2(a)–(c)**, the relaxation time, $1/R_i$, decreases by 10 times in each case so that the detonation wave changes from unsteady to steady. In **Figure 2(d)–(f)**, the shaded area, enclosed by the curve $\Delta_{2,xx}$ and the x-axis, presents a rough description on the global TNE effect around the detonation wave. It is clear that the viscosity (and/or heat conductivity) decreases the local TNE while increases the global TNE around the detonation wave.

Figure 3 shows some numerical results aiming to investigate the main mechanisms resulting in entropy increase and their relative importance in the combustion system [19]. It is clear that, in the checked cases, the most pronounced contribution to entropy increase is from the chemical reaction, Δs_{CHEM}, the lest contribution is from the non-organized energy flux, Δs_{NOEF}, the contribution of non-organized momentum flux, Δs_{NOMF}, is in between. With the increasing of Mach number, the entropy production caused by non-organized momentum flux becomes more remarkable.

3.2. Multiphase flow with phase separation

Phase separation is an important branch in the field of multiphase flows. It is also a kind of non-equilibrium phase transition. The key step for modeling phase separation is to incorporate the non-ideal gas effects into the discrete Boltzmann equation. Enskog equation can be regarded as an extension of Boltzmann equation under the hard-ball molecule model [1]. Although the specific treatments may be different, the aims are the same. Those are all to replace the equation of state of ideal gas with a more practical one.

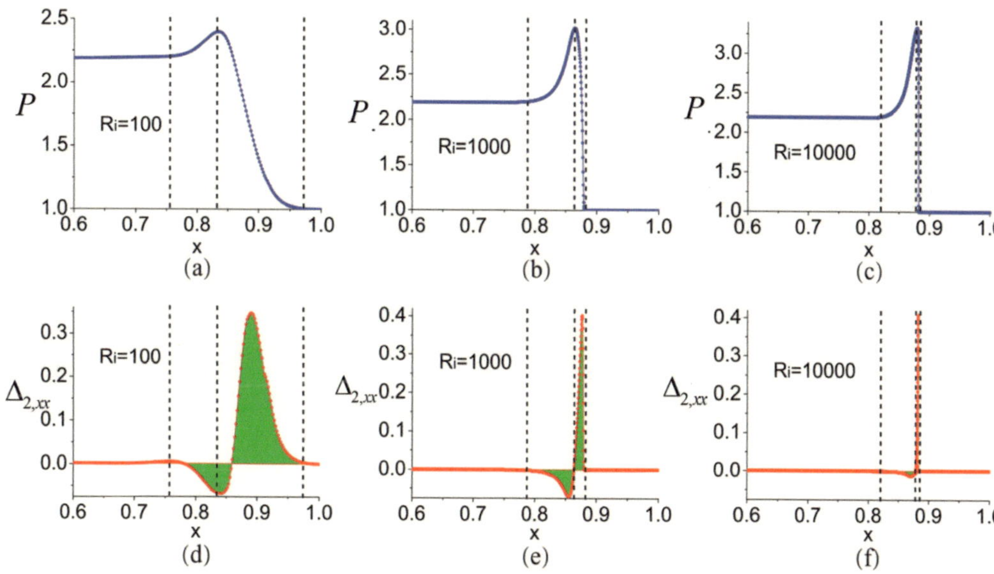

Figure 2. Profiles of physical quantities for different relaxation times. (a)–(c) The pressure P. (d)–(f) The corresponding non-equilibrium quantities $\Delta_{2,xx}$ (from Ref. [22] with permission).

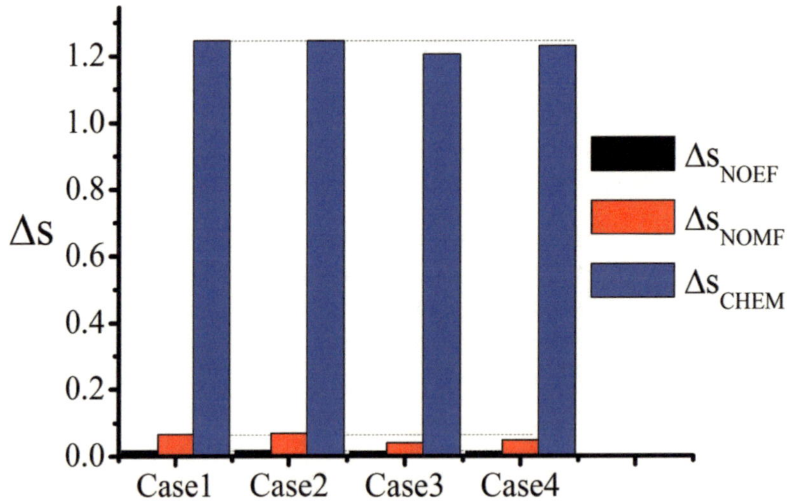

Figure 3. Mechanisms for global entropy production in four cases (from Ref. [19] with permission).

In 2007 Gonnella, Lamura, and Sofonea (GLS) [27] introduced an appropriate inter-particle interaction to the Watari's model [28] to describe van der Waals fluids. The evolution equation of GLS model reads:

$$\frac{\partial f_{ki}}{\partial t} + v_{ki\alpha}\frac{\partial f_{ki}}{\partial r_\alpha} = -\frac{1}{\tau}\left(f_{ki} - f_{ki}^{eq}\right) + I_{ki}, \tag{28}$$

where the external force term I_{ki} takes the following form:

$$I_{ki} = -\left[A + B_\alpha(v_{ki\alpha} - u_\alpha) + \left(C + C_q\right)(v_{ki\alpha} - u_\alpha)^2\right]f_{ki}^{eq} \tag{29}$$

In a recent work [20], the GLS model was further developed to be a kinetic model which can be used to access both the hydrodynamic non-equilibrium and the thermodynamic non-equilibrium. To roughly and averagely estimate the derivation amplitude from the thermodynamic equilibrium, a TNE strength can be defined as

$$D = \sqrt{\Delta_2^{*2} + \Delta_3^{*2} + \Delta_{3,1}^{*2} + \Delta_{4,2}^{*2}} \tag{30}$$

Figure 4 shows that the maximum value point can work as a physical criterion to discriminate the two stages, spinodal decomposition and domain growth, of phase separation. The TNE strength increases with time in the first stage while decreases with time in the second stage. More details are referred to the original publication [20].

3.3. Rayleigh-Taylor interfacial instability

Rayleigh-Taylor instability (RTI) occurs at the interface between two fluids with different densities. The compressible RTI system can be described by [23, 24]

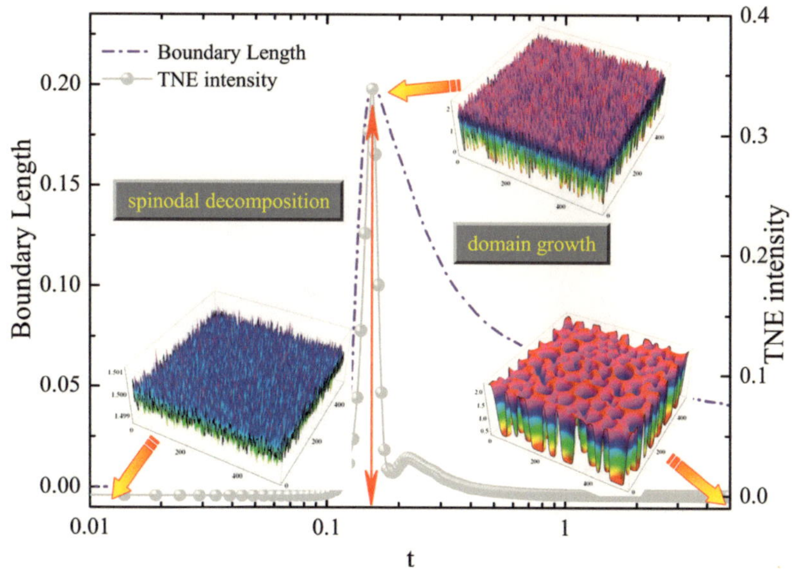

Figure 4. Evolutions of the boundary length L and the TNE strength D for the phase separation process (see Ref. [20] for more details).

$$\frac{\partial f_i}{\partial t} + v_{i\alpha}\frac{\partial f_i}{\partial r_\alpha} - \frac{a_\alpha(v_{i\alpha} - u_\alpha)}{RT} f_i^{eq} = -\frac{1}{\tau}\left(f_i - f_i^{eq}\right) \tag{31}$$

where a_α is the acceleration due to external force.

Generally, the depth of the mixing layer is an important parameter to measure the evolution of RTI. For incompressible RTI, the measurement is readily performed by tracing the constant density. However, for the compressible case, how to measure the mixing layer remains a thorny problem. Here we present two independent interface-tracking methods. One is by tracking the mean temperature of the upper and bottom fluids while the other is by tracking the maximum values of TNE characteristic quantities, such as $\Delta^*_{3.1,y}$. The latter method is based on the fact that $\Delta^*_{3.1,y}$ takes its maximum value at the position of the interface along the y direction of the spike and bubble. The perturbation amplitudes developing with time obtained by the two methods are shown in **Figure 5**. The good agreement shows that the two approaches validate each other and the local TNE, $\Delta^*_{3.1,y}$, can be used to track interfaces in numerical experiments.

3.4. Compressible flows under shock

Figure 6 shows the profiles of physical quantities in the process where the shock wave passes outwards from the heavy medium to the light one. **Figure 6(a)** and **(b)** are for the case without and with initial perturbations at the interface, respectively. From left to right, one can find three kind of interfaces, the rarefaction wave, material interface and shock wave, which are indicated by dashed lines [25].

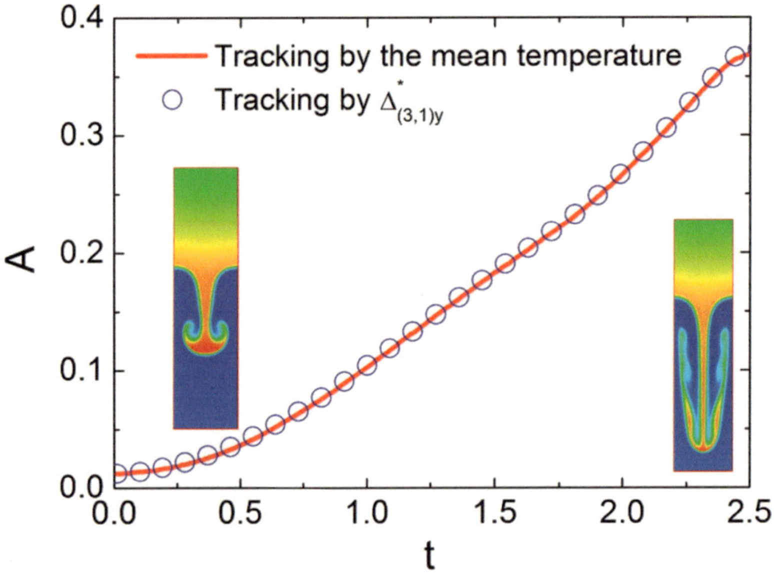

Figure 5. The perturbation amplitudes developing with time obtained by two different tracking approaches (see Ref. [23] for more details).

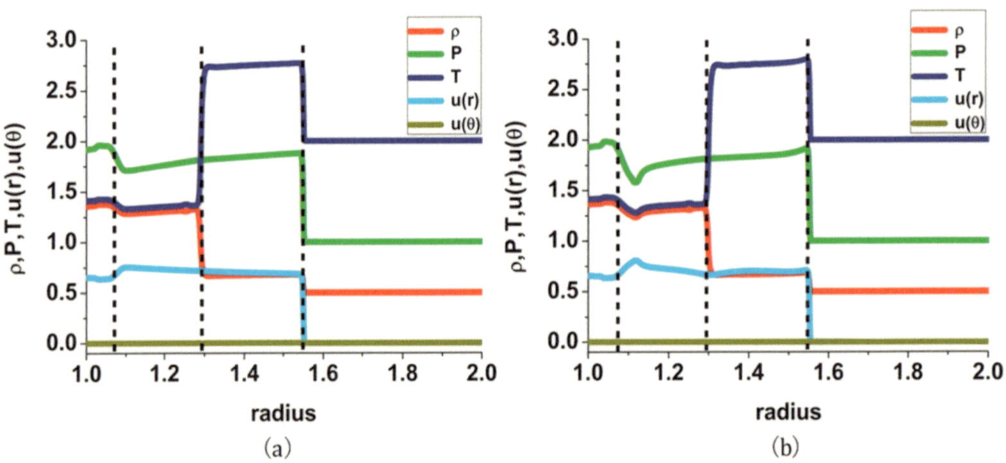

Figure 6. Profiles of physical quantities in the process of shock wave passing outwards from the heavy medium to the light one. (a) Without initial perturbations at the interface. (b) With initial sinusoidal perturbation at the interface (from Ref. [25] with permission).

According to the TNE information, the main feature of actual particle velocity distribution function can be qualitatively recovered. **Figure 7** shows an example, where the interface is not perturbed initially. The details are referred to Ref. [25]. DBM simulations [25] show that the shear stress exists only for the oblique shocking. As a consistent correspondence, MD results [29] show that fluctuating shear stresses exist if observed in a scale with a few angstroms, while their mean value becomes negligibly small when being averaged over a scale with several hundred angstroms.

A comparison of the DBM results and those of the MD is shown in **Figure 8**. **Figure 8(a)** and **(b)** shows the DBM results for the cases where the interface is not and is perturbed initially, respectively. **Figure 8(c)** shows the shear stresses from MD simulation. Since the MD uses particle description, the existence of locally fluctuating shear stress corresponds to the observation in the case with perturbed interface; the observation that mean value of shear stresses becomes negligibly small in a much larger scale roughly correspond to the case with non-perturbed interface for the DBM simulations.

3.5. Shock wave in plasma

Figure 9 shows an example for that the TNE effects can be used to physically discriminate shock wave in plasma from those in common fluid. From the first two rows, the two TNE quantities, Δ_2^* and $\Delta_{4,2}^*$, show quite similar behaviors around shock wave and/or detonation wave in common fluid, even though they two have different physical meanings. However, the two quantities show qualitative difference around shock wave in plasma [30].

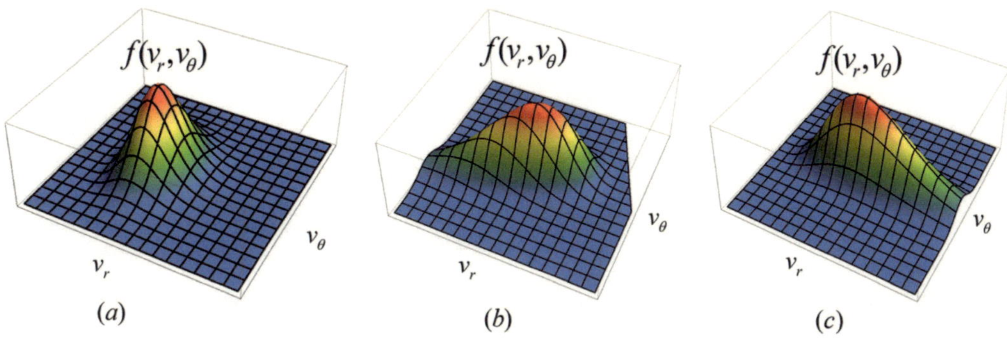

Figure 7. Sketches of the actual distribution functions in velocity space (v_r, v_θ). Panels (a)–(c) show the recovered distribution functions at the rarefaction wave, the material interface, and the shock wave, respectively (from Ref. [25] with permission).

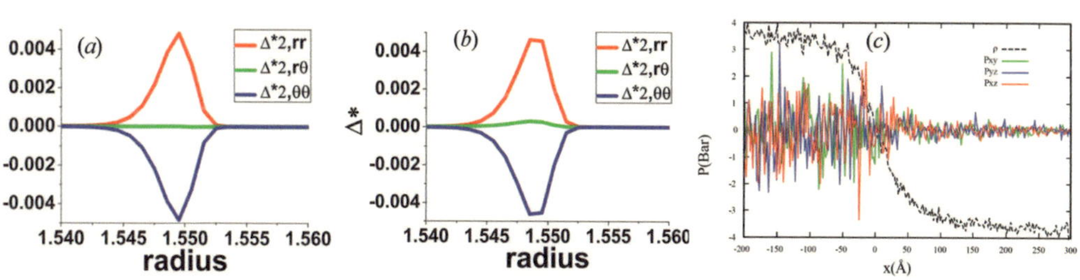

Figure 8. Shear stress within shock wave. (a) DBM results for the case without initial interface perturbation. (b) DBM results for the case with initial interface perturbation. (c) MD results (figures (a) and (b) are from Ref. [25] with permission, figure (c) is from Ref. [29] with permission).

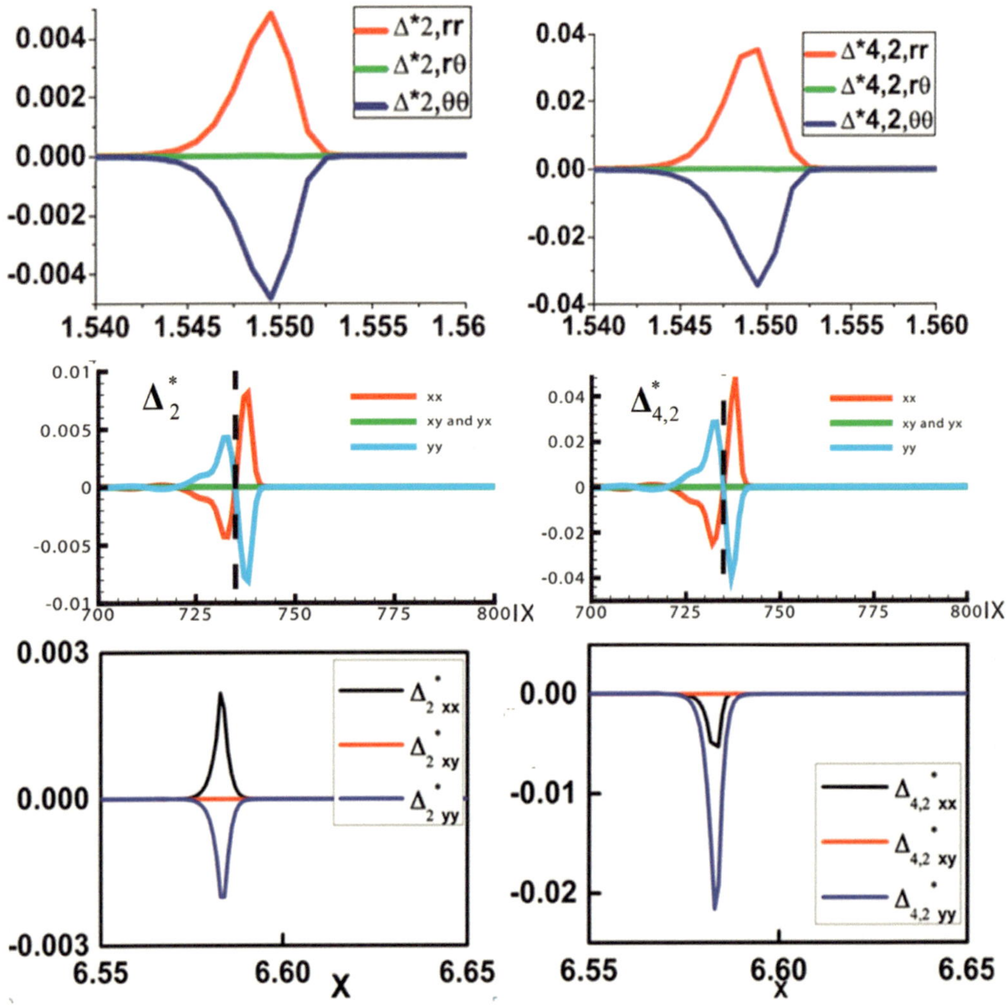

Figure 9. TNE characteristics of three types of shock waves. The three rows, from top to bottom, correspond to pure shock wave (see Ref. [25] for more details), shock wave with chemical reaction (detonation) (see Ref. [21] for more details), and shock wave in plasma (see Ref. [30] for more details). The profiles on the left column show the values of Δ_2^* around the wave fronts while the profiles on the right column show the values of $\Delta_{4,2}^*$.

4. Summary

Understanding compressible flows need more time. DBM presents a convenient way to model and simulate systems with trans-scale Knudsen number. Mathematically, the only difference of discrete Boltzmann from the traditional hydrodynamic modeling is that the NS equations are replaced by a discrete Boltzmann equation. But physically, this replacement has a significant gain: a DBM is roughly equivalent to a hydrodynamic model supplemented by a coarse-grained model of the TNE, where the hydrodynamic model can be and can also beyond the

NS. The TNE provided by DBM has been used to investigate non-equilibrium effect during detonation process, to discriminate different stages of phase separation, to recover actual particle velocity distribution function qualitatively, to track the interfaces of different fluid, and to discriminate shock wave in plasma from those in common fluid. More use of those TNE quantities are further being discovered with the deeper investigation of the compressible and complex flows.

Acknowledgements

We warmly thank Profs. Wei Kang, Zhihua Chen, Yanbiao Gan, Feng Chen, Chuandong Lin, Huilin Lai, Zhipeng Liu, Bo Yan, et al. for helpful discussions. We acknowledge support of the National Natural Science Foundation of China (under Grant Nos. 11475028 and 11772064), Science Challenge Project (under Grant No. JCKY2016212A501 and TZ2016002), and Science Foundation of Laboratory of Computational Physics.

Author details

Aiguo Xu[1,2]*, Guangcai Zhang[1] and Yudong Zhang[1,3,4]

*Address all correspondence to: ag_xu8009@163.com

1 Institute of Applied Physics and Computational Mathematics, Beijing, China

2 Center for Applied Physics and Technology, MOE Key Center for High Energy Density Physics Simulations, College of Engineering, Peking University, Beijing, China

3 Key Laboratory of Transient Physics, Nanjing University of Science and Technology, Nanjing, China

4 Graduate School, China Academy of Engineering Physics, Beijing, China

References

[1] Chapman S, Cowling TG. The Mathematical Theory of Non-uniform Gases. third ed. New York: Cambridge: University Press; 1970

[2] Rapaport DC. The Art of Molecular Dynamics Simulation. 2nd ed. New York: Cambridge: University Press; 2004

[3] Bird GA. Molecular Gas Dynamics and the Direct Simulation of Gas Flows. New York: Clarendon Press; 2003

[4] Chen S. Non-equilibrium Statistical Mechanics. Beijing: Science Press; 2010

[5] Shen H. Statistical Mechanics. Hefei: China University of Sci. & Tech. Press; 2011

[6] Xu A, Zhang G, Ying Y. Progress of discrete Boltzmann modeling and simulation of combustion system. Acta Physica Sinica. 2015;**64**(18):184701

[7] Bhatnagar PL, Gross EP, Krook M. A model for collision processes in gases. I. Small amplitude processes in charged and neutral one-component systems. Physical Review. 1954;**94**(3):511

[8] Succi S. The Lattice Boltzmann Equation for Fluid Dynamics and beyond. New York: Clarendon Press; 2001

[9] Qian YH, d'Humières D, Lallemand P. Lattice BGK models for Navier-stokes equation. EPL (Europhysics Letters). 1992;**17**(6):479

[10] Chen H, Chen S, Matthaeus WH. Recovery of the Navier-stokes equations using a lattice-gas Boltzmann method. Physical Review A. 1992;**45**(8):R5339

[11] Holway LH Jr. New statistical models for kinetic theory: Methods of construction. The. Physics of Fluids. 1966;**9**(9):1658-1673

[12] Shakhov EM. Generalization of the Krook kinetic relaxation equation. Fluid. Dynamics (Pembroke, Ont.). 1968;**3**(5):95-96

[13] Larina IN, Rykov VA. Kinetic model of the Boltzmann equation for a diatomic gas with rotational degrees of freedom. Computational Mathematics & Mathematical Physics. 2010;**50**(12):2118-2130

[14] Liu G. A method for constructing a model form for the Boltzmann equation. Physics of Fluids. 1990;**2**(2):277-280

[15] Chen F, Xu A, Zhang G, et al. Two-dimensional MRT LB model for compressible and incompressible flows. Frontiers of Physics. 2014;**9**(2):246-254

[16] Xu A, Zhang G, Gan Y, et al. Lattice Boltzmann modeling and simulation of compressible flows. Frontiers of Physics. 2012;**7**(5):582-600

[17] Xu A, Zhang G, Li Y, et al. Modeling and simulation of nonequilibrium and multiphase complex systems: Lattice Boltzmann kinetic theory and application. Progress in Physics. 2014;**34**(3):136-167

[18] Xu A, Zhang G, Gan Y. Progress in studies on discrete Boltzmann modeling of phase separationprocess. Mechanics in Engineering. 2016;**38**(4):361-374

[19] Zhang Y, Xu A, Zhang G, et al. Kinetic modeling of detonation and effects of negative temperature coefficient. Combustion & Flame. 2016;**173**:483-492

[20] Gan Y, Xu A, Zhang G, et al. Discrete Boltzmann modeling of multiphase flows: Hydrodynamic and thermodynamic non-equilibrium effects. Soft Matter. 2015;**11**(26):5336

[21] Yan B, Xu A, Zhang G, et al. Lattice Boltzmann model for combustion and detonation. Frontiers of Physics. 2013;**8**(1):94-110

[22] Xu A, Lin C, Zhang G, et al. Multiple-relaxation-time lattice Boltzmann kinetic model for combustion. Physical Review E Statistical Nonlinear & Soft Matter Physics. 2015;**91**(4): 043306

[23] Lai H, Xu A, Zhang G, et al. Nonequilibriumthermohydrodynamic effects on the Rayleigh-Taylor instability in compressible flows. Physical Review E. 2016;**94**(1):023106

[24] Chen F, Xu A, Zhang G. Viscosity, heat conductivity, and Prandtl number effects in the Rayleigh–Taylor instability. Frontiers of Physics. 2016;**11**(6):114703

[25] Lin C, Xu A, Zhang G, et al. Polar-coordinate lattice Boltzmann modeling of compressible flows. Physical Review E Statistical Nonlinear & Soft Matter Physics. 2014;**89**(1):013307

[26] Ju Y. Recent progress and challenges in fundamental combustion research. Advances in Mechanics. 2014;**44**(1):26-97

[27] Gonnella G, Lamura A, Sofonea V. Lattice Boltzmann simulation of thermal nonideal fluids. Physical Review E. 2007;**76**(3):036703

[28] Watari M, Tsutahara M. Two-dimensional thermal model of the finite-difference lattice Boltzmann method with high spatial isotropy. Physical Review E Statistical Nonlinear & Soft Matter Physics. 2003;**67**(2):036306

[29] Liu H. Shock waves and hydrodynamic instabilities under the condition of inertial confinement fusion. [Ph.D thesis]. Beijing: Peking University; 2017

[30] Liu Z. Non-equilibrium Characteristic of Shock Waves in Plasma, Post-Doctoral Research Report. Beijing: Institute of Applied Physics and Computational Mathematics; 2017

Chapter 3

Collective Mode Interactions in Lorentzian Space Plasma

Nazish Rubab and Sadia Zaheer

Additional information is available at the end of the chapter

http://dx.doi.org/10.5772/intechopen.71847

Abstract

Plasmas exhibit a vast variety of waves and oscillations in which moving charged particle produce fields which ultimately give rise to particle motion. These wave-particle effects are used in the acceleration heating methods of plasma particles, and in wave generation as well. Plasmas are often manipulated with EM waves, e.g., Alfvén waves are long-wavelength modes (drift-waves) where fluid theory is most reliable, while for short wavelength modes (e.g., Kinetic Alfvén waves), collisionless effects becomes important. In this chapter, the properties of kinetic Alfvén waves are aimed to study by employing two potential theory by taking particle streaming and Weibel instability with temperature anisotropy in a Lorentzian plasma.

Keywords: KAWs, Lorentzian distribution, streaming and temperature anisotropy, dusty plasma

1. Introduction

This chapter addresses one of the intriguing topics of Astrophysics—the existence of kinetic Alfvén wave (KAW) and the important consequences for astrophysical and space science to explore and investigate the new avenues. Due to the fact that KAWs have non-zero electric field $E_{\|}$ which is parallel to background magnetic field and possess anisotropic polarized and spatial structures which contribute to particle energization. It is an interesting mechanism that KAWs can accelerate the field-aligned charged particles and has been applied in the dissipation of solar wind turbulence, the acceleration, and heating of charged particles in both the filed-aligned and perpendicular directions and is anticipated to play a vital role in the particle energization in laboratory, space and astrophysical plasma. The progress reported here would have immense impact and hence a small step in particular direction.

The solar wind plasma is hot and weakly collisional, existing in a state far from thermal equilibrium [1] as observed in situ in the solar wind through its nonthermal characteristics of velocity distribution function (VDF). The electron VDFs measured at 1 AU have been used as boundary condition to determine the VDFs at different altitudes. It has been confirmed that for several solar radii, the suprathermal population of particles is present in the corona [2]. For low collision rates in such plasmas, the particles can develop temperature anisotropy and the VDFs become slanted and build up high energy tails and heat fluxes along the magnetic field direction especially in fast winds and energetic interplanetary shocks. Various processes in a collisionless solar wind plasma lead to the development of particle temperature anisotropy to generate plasma instabilities which are often kinetic in nature. The free energy sources associated with the deviation from the thermodynamical equilibrium distribution function could also excite plasma waves [3–11].

In general, the study of plasma waves and micro-instabilities in the solar wind shows that proton VDFs are prone to anisotropic instability and originate to be stable or marginally stable. Marsch [12] has discussed four significant electrostatic and electromagnetic wave modes and free energy sources to make them unstable. For example, the electrostatic ion acoustic wave may be destabilized by the ion beams and electrons and electron heat flux, [13] the electromagnetic ion Alfvén-cyclotron wave needs proton beam and temperature anisotropy, magnetosonic wave requires proton beam and ion differential streaming and whistler-mode and lower-hybrid wave [14] unstable solutions. Among several electromagnetic instabilities, the kinetic Alfvén wave instability is the most important one.

The satellite missions in space and astrophysical plasmas have confirmed the presence of non-Maxwellian high energy and velocity tails in the particle distribution function and found in the magnetosphere of Saturn, Mercury, Uranus and Earth [2, 15–17]. The non-Maxwellian distribution of charged particles has been observed to give a better fit to the thermal and superthermal part by employing kappa distribution, since it fits both thermal and suprathermal parts in the energy velocity spectra.

The subject area of this chapter involves the basic research of space plasma physics and in particular, focuses the investigations of electrostatic and electromagnetic waves in a multi-component dusty (complex) Maxwellian and non-Maxwellian plasmas. In the last few years, various power-law distribution functions (in velocity space), i.e., kappa and (r, q) have been used to investigate collective phenomena and associated instabilities, such as dust-acoustic waves, kinetic Alfvén waves, Weibel instabilities, dust charging processes (in linear and nonlinear regimes) in space and astrophysical situations for better fitting the observational data in comparison to Maxwell distribution. These distributions have relevance to space plasmas containing solar wind, interstellar medium, ionosphere, magnetosphere, auroral zones, mesosphere, lower thermosphere, etc.

When the intense radiations interact with plasmas, it ends up with many applications like instabilities, inertial confinement fusion [18], and pulsar emissions [19]. These instabilities further generate turbulent electromagnetic fields in plasma regimes. We can characterize instabilities as electrostatic as well as electromagnetic according to the conditions provided by nature [20]. In this chapter, we shall also discuss electromagnetic instability called Weibel instability in a Lorentzian plasma. The free energy source available for Weibel instability is temperature anisotropy and can be developed in magnetically confined and magnetic free plasma environment as well. First time Weibel [21] came up with the calculations of impulsive growing transverse waves with anisotropic velocity distribution function in 1958. This instability developed when the electrons in the fluctuating magnetic field generates momentum flux, this flux sequentially effects velocity $< \mathbf{v} >$ (and ultimately current density $< \mathbf{J} >$) as to increase the fluctuating field [22]. The property of Weibel instability is that it is different from normal resonant wave-particle instabilities because it depends on effects in bulk plasma without any resonant particle contributions [23]. The particle distribution functions in kinetic model adequately describe a physical phenomenon in terms of time and phase space configurations providing more information to investigate plasma waves, instabilities, plasma equilibrium, Landau damping phenomena, etc. In this chapter, we shall review the kinetic/Inertial Alfvén waves and instabilities, the effects of dust grain charging as well as field aligned/cross field currents, streaming velocity and the non-Maxwellian power-law distribution and its effect on various electromagnetic modes. We intend to show that the presence of dust grains introduces a new cutoff frequency Ω_{dlh} which is associated with the motion of mobile charged particles. Moreover, an interesting feature is to show that the employed model inhibits the temperature anisotropy and supports the velocity anisotropy. Further, we shall calculate the linear dispersion relation for Weibel instability in Lorentzian plasma ($B_0 = 0, B_0 \neq 0$) by using linearized, nonrelativistic Vlasov equation. We shall solve $Z^*_\kappa(\alpha)$ by assuming $\alpha < 1$ or $\alpha > 1$ for $\kappa = 3, 5, 7$.

2. Model and methodology

In long-wavelength modes the fluid theory is most reliable, while for short wavelength modes (like KAWs), collisionless effects are important, for example, Landau damping due to finite ion Larmor radius explains observed damping rate and in dusty plasmas and charge fluctuations. Kinetic Alfvén waves (KAWs) are small scale dispersive Alfvén waves (AWs) which plays a significant role in particle acceleration and plasma heating. A coupling mechanism between small-scale KAWs and large-scale AWs in the presence of superthermal particles has been discussed which in turns giving rise to the excitation of KAWS in a solar/stellar wind plasma have been studied in the past. In this chapter, we intend to show the relationship between the growth rates of excited anisotropic KAWs and perpendicular wavelength by taking charge fluctuation and Landau damping variations into account. Moreover, when the perpendicular component of the wavelength, when comparable to the ion gyroradius, a magnetic field aligned electric field plays a significant role in the plasma acceleration/heating. Utilizing a two potential theory along with kinetic description, the properties of kinetic Alfvén waves are aimed to investigate different modes in low beta plasmas by incorporating the streaming effects. We present overview of electromagnetic KAW streaming instability in a collisionless dusty magnetoplasma, whose constituents are the electrons, ions and negatively charged dust particles. The interaction between monochromatic electron/ion beam with plasma is also discussed under various conditions. Further, to calculate the linear dispersion relation for Weibel instability in unmagnetized Lorentzian plasma, we shall employ linearized, nonrelativistic Vlasov equation.

2.1. Two potential theory

In a low beta plasma, $\beta < 1$, the electric field can be described by two potential theory or fields expressing the electromagnetic perturbations with shear perturbations only in the magnetic field. We may neglect the electromagnetic wave compression along the direction of magnetic field ($B_{1z} = 0$), which leads to the coupling of Alfvén-acoustic mode. Thus, we adopt a two potential theory which represents both the transverse and parallel components of the electric field as $E_\perp = -\nabla_\perp\varphi$ and $E_\| = -\partial\psi/\partial z$, with $\varphi \neq \psi$ [23]. We shall also consider the charge on the dust grain which may fluctuate according to the plasma conditions. At equilibrium, the charge neutrality imposes the condition $n_{e0} - n_{i0} + Z_{d0}n_{d0} = 0$, where $n_{e0}(n_{i0})$ is the electron (ion) number density and Z_{d0} is the equilibrium dust charging state.

The linearized Poisson and Maxwell equations in terms of parallel and perpendicular operators can be expressed as

$$\nabla_\perp^2\varphi + \partial_z^2\psi = \frac{1}{\epsilon_0}(n_{e1} + Z_{d0}n_{d1} + n_{d0}Z_{d1} - n_{i1}), \tag{1}$$

and

$$\partial_z\nabla_\perp^2(\varphi - \psi) = \mu_0\partial_t(J_{e1z} + J_{i1z} + J_{d1z}), \tag{2}$$

where $\epsilon_0(\mu_0)$ is the permittivity (permeability) of the free space and J_{j1z} represents the field aligned current density for jth species ($j = e$ for electrons, i for ions and d for dust grains). In obtaining Eq. (2), we have ignored the factor $\nabla_\perp\nabla_z E_z - \nabla^2 E_\perp$. The main idea is to decouple the compressional Alfvén mode under the assumption $(\nabla \times E)_z = -(\partial B/\partial t)_z = 0$, i.e., to highlight bending of line of force and minimize any change in field strength due to wave compression. Moreover, $(\nabla \times E)_z$ engross only $E_\perp$ and the least restrictive assumption for $(\nabla \times E)_z$ to vanish is $E_\perp = -\nabla_\perp\varphi$ in which the perpendicular electric field $E_\perp$ is electrostatic, leaving an incompressible mode. When $\varphi = \psi$, the twist of the magnetic field lines vanishes, therefore, the incompressible shear modes have $\nabla.u_\perp = 0 = B_{1z}$, and $E_\perp = -\nabla_\perp\varphi$.

$$(\partial_t + \mathbf{v}\cdot\nabla)f_{j1} + \frac{q_j}{m_j}(\mathbf{E} + \mathbf{v}\times\mathbf{B}).\nabla_\mathbf{v}\, f_{j0} = 0, \tag{3}$$

$$f_{j1} = \frac{q_j k_z \psi}{m_j(\omega - k_z v_z)}\frac{\partial f_{j0}}{\partial v_z}, \tag{4}$$

where f_{j0} is the equilibrium distribution function. The dynamics of cold and magnetized dust is governed by set of fluid equations, i.e.,

$$\partial_t v_d = \frac{Z_d e}{m_d}\mathbf{E} + \mathbf{V}_d \times \boldsymbol{\omega}_{cd} \tag{5}$$

and

$$\partial_t n_d + div(n_d\mathbf{V}_d) = 0 \tag{6}$$

2.2. Number density and current density perturbations

Here, we may define the number density and current density as

$$\begin{aligned} n_{j1} &= n_0 \int f_{j1} d\mathbf{v}, \qquad j = e, i, d \\ \mathbf{J}_{j1} &= q_j n_0 \int \mathbf{v} f_{j1} d\mathbf{v}, \end{aligned} \tag{7}$$

3. Dispersion and damping of kinetic Alfvén waves (KAW)

Kinetic Alfvén waves (KAWs) are small scale dispersive Alfvén waves (AWs) which plays a significant role in particle acceleration, turbulence, wave particle interaction and plasma heating. Kinetic processes prevail in the regimes where plasma is dilute, multi-component, and non-uniform. A coupling mechanism between small-scale KAWs and large-scale AWs with superthermal plasma species which in turns gives rise to the excitation of KAWS in a solar/stellar wind plasma has proved dispersive Alfvén waves responsible for the solar wind turbulence especially when the turbulence cascade of these electromagnetic waves transfer from larger to smaller scale as compared to proton gyro radius. Moreover, from spacecraft observations in ionospheric plasma, it is evident that Alfvénic Poynting flux is responsible to transfer the energy for particle acceleration. All the energized auroral particles accelerate in ionosphere, initiate Joule heating phenomenon and stream out into the magnetosphere [25–28].

There are number of studies to show the relationship between the growth rates of excited anisotropic KAWs and perpendicular wavelength by taking charge fluctuation and Landau damping variations into account. Moreover, the perpendicular component of wavelength, when comparable to ion gyroradius, a magnetic field aligned electric field plays a significant role in plasma acceleration/heating.

One of the important features in astrophysical plasma is the transportation of electromagnetic energy through the wave interaction with thermal plasma ions [29–31]. The KAW plays a vital role to transfer the wave energy through Landau damping (when thermal electrons travel along the magnetic field lines), which is regarded as collisionless damping of low-frequency waves and during this process the particles gain kinetic energy from the wave. This process can only happen when the distribution function has a negative slope which results in the heating of plasmas or acceleration of electrons along the magnetic field direction [24]. Recent studies also suggest the impact of non-Maxwellian distribution functions on the dynamics of solar wind and auroral plasma [32]. This study shows that the plateau formation in the parallel electron distribution functions minimize the Landau damping rate significantly.

In this chapter, the properties of kinetic Alfvén waves would be discussed by employing two potential theory, Maxwell equations and Vlasov model to study different plasma modes and by taking streaming of charged particles along and across the field direction in a Maxwellian and Lorentzian plasma.

3.1. Kinetic Alfvén waves in Maxwellian plasma

The propagation of kinetic Alfvén waves in a dusty plasma with finite Larmor radius effects will be discussed using a fluid-kinetic formulation by taking charge variations of dust particles. The coupling of Alfvén-acoustic mode results in the formations of kinetic Alfvén wave which would be discussed in forth coming subsections. In a magnetized plasma, we shall consider the electrons are thermal and strongly magnetized obeying an equilibrium Maxwellian distribution, while ions are hot and magnetized so that finite Larmor radius can be taken into account. For ions, we may employ Vlasov equation by utilizing guiding center approach to obtain the perturbed distribution function for an electromagnetic wave when the electric field and the wave vector $\mathbf{k}$ lie in the xz plane and, $\mathbf{B}_0 = (0, 0, \mathbf{B}_0)$, $\mathbf{k} = (k_\perp, 0, k_z)$,

$$f_{i1} = -\left(\frac{n_{i0}e}{T_i}\right)\sum_l\sum_n \frac{k_z v_z \psi + n\Omega_{ci}\varphi}{\omega - n\Omega_{ci} - k_z v_z}\exp\left(i(n-l)\theta\right)J_n\left(\frac{k_\perp v_\perp}{\Omega_{ci}}\right)J_l\left(\frac{k_\perp v_\perp}{\Omega_{ci}}\right)f_{i0}, \tag{8}$$

where J_n is the Bessel function of first kind, having order n and f_{i0} is the equilibrium distribution function. On using Eq. (7), we obtain the modified number and current densities for hot and magnetized ions and thermal electrons, i.e.,

$$n_{i1} = -\frac{en_{i0}}{k_z m_i v_{ti}^2}\sum_n \left[k_z v_{ti}\psi(1 + \xi_{in}Z(\xi_{in})) + n\Omega_{ci}\varphi Z(\xi_{in})\right]I_n(\vartheta_i)e^{-\vartheta_i}, \tag{9}$$

$$J_{i1z} = -\frac{n_{i0}e^2}{T_i k_z}\sum_n \left[(1 + \xi_{in}Z(\xi_{in}))(k_z v_{ti}\xi_{in}\psi + n\Omega_{ci}\varphi)\right]I_n(\vartheta_i)e^{-\vartheta_i}, \tag{10}$$

and

$$n_{e1} = \frac{en_{e0}}{T_e}\psi(1 + \xi_e Z(\xi_e)), \tag{11}$$

$$J_{e1} = \frac{e^2 n_{e0}\psi}{m_e v_{te}}\xi_e Z'(\xi_e), \tag{12}$$

where I_n is the modified Bessel function with argument $\vartheta_{i,e} = k_\perp v_{ti,e}^2/2\Omega_{ci,e}$ and $Z(\xi_{in})$ is the usual dispersion function for a Maxwellian plasma with $\xi_{in} = (\omega - n\Omega_{ci})/k_z v_{ti}$ and Z' is the derivative of Z with respect to its argument.

The dust component is considered to be cold and unmagnetized such that $\omega << \omega_{cj}$, $k_z v_{te} >> \omega$ and $k_z v_{ti} << \omega$, therefore, we use hydrodynamical model with momentum balance equation and continuity equation For cold and unmagnetized dust and thus we obtain

$$n_{d1} = -\frac{n_{d0}Z_{d0}e}{m_d\omega^2}\left[k_\perp^2\varphi + k_z^2\psi\right], \tag{13}$$

and

$$J_{d1z} = \frac{n_{d0}Q_{d0}^2}{m_d\omega}k_z\psi. \tag{14}$$

To find the relation between φ and ψ, the expressions of n_{e1}, n_{i1} and n_{d1} are used into Eq. (1) and J_{i1z}, J_{e1z}, and J_{d1z} into Eq. (2) to obtain the following coupled equations:

http://dx.doi.org/10.5772/intechopen.71847

$$
\begin{aligned}
A\varphi + B\psi &= 0,\\
C\varphi + D\psi &= 0,
\end{aligned}
\tag{15}
$$

The coefficients in Eq. (15) are given by

$$
\begin{aligned}
A &= \frac{k_\perp F_i}{\omega^2}\omega'\left(1-\frac{3}{4}\vartheta_i\right),\\
B &= \lambda_{De}^{-2}-\left(k_z^2\omega_{pd}^2\right)/\omega^2,\\
C &= c^2k_z^2k_\perp^2,\\
D &= \frac{1}{k_z}\frac{2\omega^2}{\lambda_{De}^2}-k_z\varepsilon_\perp,
\end{aligned}
\tag{16}
$$

where $F_i = \omega_{pi}^2/\omega_{ci}^2$, $\omega' = \omega^2 - \Omega_{(dlh)i}^2$, $\varepsilon_\perp = \omega_{pd}^2 + c^2k_\perp^2$ and $\Omega_{(dlh)i}^2 = \omega_{pd}^2/F_i$. The solution of homogeneous Eq. (15) in the form of a biquadratic equation, i.e.,

$$
p\omega^4 + Q\omega^2 + R = 0 \tag{17}
$$

where,

$$
\begin{aligned}
p &= \frac{2\omega_{pe}^2}{k_z v_{te}^2},\\
Q &= -\frac{2\omega_{pd}^2}{k_z v_{te}^2}\Omega_{(dlh)i}^2 - k_z\epsilon_\perp - \frac{k_z v_{Ai}^2}{\lambda_{De}^2}\left(1+\frac{3}{4}\vartheta_i\right),\\
R &= \Omega_{(dlh)i}^2 k_z\epsilon_\perp + k_z^3 v_{Ai}^2\omega_{pd}^2\left(1+\frac{3}{4}\vartheta_i\right)
\end{aligned}
\tag{18}
$$

where $V_{Ai} = c\omega_{ci}/\omega_{pi}$ is the Alfvén velocity of ions. The solution of biquadratic equation in the form of kinetic Alfvén wave is as follows,

$$
\omega^2 = \Omega_{(dlh)i}^2 + k_z^2V_{Ai}^2\left[1+\left\{\frac{3}{4}+T'\Lambda_i\frac{\epsilon_\perp}{c^2k_\perp^2}\right\}\vartheta_i,\right] \tag{19}
$$

where, $T' = \frac{T_e}{T_i}$. and $\Lambda_i = n_{i0}/n_{e0}$ This shows the dispersion relation of kinetic Alfvén waves in the presence of mobile dust that are the extension of shear Alfvén waves in the range of small perpendicular wavelength. The first term on the R. H. S appears due to dust dynamics, i.e., a new cut off frequency due to the hybrid dynamics of cold dust and magnetized ions which provides a limit to the propagation of electromagnetic wave. In a dustless plasma, i.e., $\omega_{pd} = 0$, we obtain usual dispersion relation in electron-ion plasma. Expressing ω in terms of real and imaginary part, $\omega = \omega_r + i\gamma$, with $\omega_r >> \gamma$, we either obtain growth or damping of KAW satisfying the condition, $\omega/k_z = v_A \le v_z$ through wave particle interactions [33, 34]. In a dusty plasma with dust charge fluctuation effect, the main mechanism of wave damping is associated with dust charge fluctuation effects as compared to Landau damping [34]. It is a well-known fact that if the particle thermal velocity exceeds the Alfvén velocity, then the particles interact with Alfvén wave as the result of wave particle interaction/resonance, the linear Landau damping prevails. In a dusty plasma, the massive dust grains move slowly as compared to Alfvén velocity, therefore they may interact with Alfvén wave through linear Landau damping (which is negligible in case of dust species) or charge fluctuation effects.

3.2. Lorentzian distribution function

A number of processes in a space based plasma lead to the development of particle anisotropy through streaming or temperature and are responsible for plasma instabilities in collision-free plasma which are frequently kinetic in nature and their persistent features have been confirmed by many spacecraft measurements, e.g., the electron energy spectra and the near-earth environment observations have witnessed the presence of superthermal populations. It is a well-known fact that the equilibrium Maxwell-Boltzmann distributions are associated with the Boltzmann collision term, but on the large scale Fokker-Plank model is not appropriate due to strong interaction and correlation in a collisionless plasma. The kinetic foundations of generalized Lorentzian statistical mechanics has been remarkably established by [35] with the generalization of Boltzmann collision term that is not based on binary collisions. The long range correlation between particles vindicates that power law distributions posses a particular thermodynamical equilibrium state. The mathematical form on isotropic Lorentzian distribution function is given by

$$f_{j0}^{\kappa} = A_{\kappa}\left[1 + \frac{1}{\kappa v_{tj\kappa}^2}\left(v_z^2 + {v_\perp}^2\right)\right]^{-\kappa-1} ;\kappa > 3/2, \tag{20}$$

where $A_\kappa = n_{i0}\left(\frac{1}{\pi\kappa v_{tj\kappa}^2}\right)^{\frac{3}{2}} \frac{\Gamma(\kappa+1)}{\Gamma(\kappa-1/2)}$

Due to the stated fact, the deviation from the Maxwellian equilibrium distribution function could also excite plasma waves by using free energy sources. Such distributions are frequently observed in solar and terrestrial environments and can be represented by anisotropy in temperature and velocity, i.e., [36]

$$f_{j\kappa}(v_z, v_\perp) = A_\kappa\left[1 + \frac{1}{\kappa}\left(\frac{(v_z - v_0)^2}{v_{tj\kappa z}^2} + \frac{(v_\perp - v_0)^2}{v_{tj\kappa\perp}^2}\right)\right]^{-\kappa-1}, \tag{21}$$

where $v_{tj\kappa}^2 = \left(\frac{2\kappa-3}{\kappa}\right)v_{tj}$, $v_{tj} = \left(\sqrt{k_B T/m_j}\right)$ is the thermal speed of jth plasma component, the number densities are represented by n and anisotropic temperatures components are represented as moment of second order

$$T_z = \frac{m}{n_c k_B}\int f_\kappa v_z^2 d^3v, \qquad T_\perp = \frac{m}{n_c k_B}\int f_\kappa v_\perp^2 d^3v, \tag{22}$$

In the limit $\kappa \rightarrow \infty$, the bi-Lorentzian function is reduced to bi-Maxwellian, $f_\kappa(v) \rightarrow f_M(v)$.

3.3. Lorentzian current and number density perturbations

Many space and astrophysical plasmas have been found to have generalized Lorentzian particle distribution functions. It is of some interest to observe the impact of the high energy tail on the current and number densities of plasma species. By using Eqs. (4), (7) and (20), we get the modified expressions of number and current densities based on kappa distribution function, i.e.,

$$n_{j1} = \pm \frac{2e\psi n_{j0}}{m_j v_{tj\kappa}^2} \left[\kappa' + \xi_{j0} Z_\kappa(\xi_{j0})\right], \tag{23}$$

and

$$J_{j1z} = -\frac{2e^2\psi n_{j0}}{m_j v_{tj\kappa}} \left[\kappa' \xi_{j0} + \xi_{j0}^2 Z(\xi_{j0})\right], \tag{24}$$

where $Z_\kappa(\xi_{j0}) = \frac{1}{\pi^{1/2}} \frac{\Gamma(\kappa+1)}{\kappa^{3/2}\Gamma(\kappa-1/2)} \int_{-\infty}^{+\infty} \frac{x dx}{(x-\xi_{j0})(1+x^2/\kappa)^{\kappa+1}}$, is the plasma dispersion function and $\kappa' = (2\kappa - 1)/2\kappa$..

3.4. KAW and instability in Lorentzian plasma

In a low β plasma, the kinetic Alfvén wave instability driven by field aligned currents has dependence on plasma β and streaming velocity of current carrying species which can be responsible for particle energization. In this subsection, we extend the above scenario of electromagnetic kinetic Alfvén wave by introducing the streaming of Lorentzian ions along an external magnetic field $(\mathbf{B}_0\hat{z})$ with constant ion drift velocity $(V_0 \| \mathbf{B}_0)$, strongly magnetized and hot electrons to be Maxwellian and cold unmagnetized dust. The plasma beta β_e is assumed to be very small. The electric field and the wave vector $\mathbf{k}$ lie in the xz plane, i.e., $\mathbf{B}_0 = (0, 0, \mathbf{B}_0\hat{z})$, $\mathbf{V}_0 = (0, 0, V_0\hat{z})$, $\mathbf{k} = (k_\perp\hat{x}, 0, k_z\hat{z})$. We again solve the Vlasov equation For hot and magnetized electrons [33] to get the number and current density of electrons as obtained in the previous section. Making use of Eqs. (4), (7) and (21), we get the perturbed number density of Lorentzian type streaming ions,

$$n_{i1} = -\frac{2en_{i0}\psi}{m_i v_{ti\kappa}^2} [\kappa' + \eta Z_\kappa(\eta)], \tag{25}$$

The longitudinal components of current density perturbation [7, 19, 37] is given by

$$J_{i1z} = -\frac{2e^2 n_{i0}\psi}{m_i v_{ti\kappa}} \left[\kappa'\eta + \eta^2 Z_\kappa(\eta)\right], \tag{26}$$

where $\eta = (\omega - k_z V_0)/k_z v_{ti\kappa}$.

By incorporating the values of n_{i1} and J_{i1} in Eqs. (1), (2) and using (15), the dispersion relation of KAW streaming instability in a Lorentzian dusty plasma is obtained as

$$1 + \frac{2\omega_{pe}^2}{k_z^2 v_{te}^2}\delta_1 + \frac{2\omega_{pi}^2}{k_z^2 v_{ti\kappa}^2}\delta_2 + \frac{\omega_{pd}^2}{k_z^2 c^2} + K - (1+K)\frac{\omega_{pd}^2}{\omega^2} = 0, \tag{27}$$

where $K = k_\perp^2/k_z^2$ and

$$\begin{aligned}\delta_1 &= \sum_n \left[\epsilon_\perp \frac{n\Omega_{ce}}{k_\perp^2 c^2 k_z v_{te}} Z(\xi_{en}) I_n e^{-\vartheta_e} + \left(1 - \frac{\omega v_{te}\xi_{en}}{c^2 k_z} + \frac{n\omega\Omega_{ce}}{k_\perp^2 c^2}\right)(1 + \xi_{en} Z(\xi_{en})) I_n e^{-\vartheta_e}\right] \\ \delta_2 &= \left(1 - \frac{2v_{ti\kappa}\omega\eta}{c^2 k_z}\right)\left(1 - \frac{\omega_{pd}^2}{\omega^2}\right)(\kappa' + \eta Z_\kappa(\eta)).\end{aligned} \tag{28}$$

A visible modification can be noticed by the effect of superthermality via the kappa-modified plasma dispersion function and the appearance of dust lower hybrid frequency due to dust effects on the dispersion characteristics. Numerous standard wave modes can originate from the above dispersion equation by applying particular limits, i.e.,

(i) $(\mathbf{k}\|\mathbf{B}_0, V_0):$ For $n = 0,\ \vartheta_e \ll 1,$ a dispersion relation two stream instability (TSI) in unmagnetized plasma is obtained [37], i.e.,

$$1 + \frac{2\omega_{pe}^2}{k_z^2 v_{te}^2}(1 + \xi_{en} Z(\xi_{en})) + \frac{2\omega_{pi}^2}{k_z^2 v_{ti\kappa}^2}(\kappa' + \eta Z_\kappa(\eta)) - \frac{\omega_{pd}^2}{\omega^2} = 0. \tag{29}$$

In the limit $\kappa \to \infty$, our results approach to its classical Maxwellian counterpart in a dustless plasma environment [38].

(ii) $(\mathbf{k}\|\mathbf{B}_0), V_0 = 0,\ (\Omega_{ci} \ll \omega \ll \Omega_{ce}):$ In a dustless plasma, we get whistler-like mode whose frequency is below the electron cyclotron frequency, i.e.,

$$\omega = k_z v_{ph},$$

where $v_{ph} = c^2 k_z \Omega_{ce}/\omega_{pe}^2$ is the phase velocity of whistler waves which is obviously not susceptible to the Lorentzian index κ. Again, in the limiting case $\omega \ll \Omega_{ce}, \vartheta_e \ll 1,$ and expanding plasma dispersion function Eq. (15) depicts the coupling of electromagnetic and electrostatic mode, i.e., shear Alfvén-acoustic mode due to thermal kinetic effects due to which shear Alfvén wave builds a longitudinal component, e.g.,

$$\left(\omega^2 - \Omega_{(dlh)e}^2\right)\left[\kappa' \frac{\omega_{pi}^2}{k_z v_{ti\kappa}^2}(\omega^2 - \omega k_z V_0) - k_z \epsilon_\perp\right] - \frac{k_z V_{Ae}^2}{\lambda_{Di\kappa}^2}\left[\omega^2 - k_z^2 \omega_{pd}^2 \lambda_{Di\kappa}^2\right] = 0, \tag{30}$$

where $\Omega_{(dlh)e}^2 = \omega_{pd}^2/F_e,\ F_e = \omega_{pe}^2/\Omega_{ce}^2,\ \lambda_{Di\kappa} = \sqrt{\frac{T_i}{4\pi n_{i0} e^2}\frac{1}{c_\kappa}}$ and $c_\kappa = (2\kappa - 1)/(2\kappa - 3)$. In the limit $\omega^2/k_z^2 V_A^2 \to 0,\ V_0 = 0,\ k_\perp^2 \rho_e^2 \ll 1$ we get the dispersion relation of Lorentzian dust-acoustic waves, $\omega^2 = (k_z^2 C_D^2)/c_\kappa,$ where $C_D^2 = Z_{d0} T_i/m_d$ and $V_{Ae} = B_0/(4\pi n_{e0} m_e)^{1/2}$ is the electron Alfvén speed with electron mass density. For a low beta plasma, the coupling between dust-acoustic and shear Alfvén wave becomes weak and two modes would decouple. In the limit $\kappa \to \infty$, we approach to a Maxwellian DAW [40]. It is worthy to mention here that due to the contribution of Lorentzian particles, the KAW instability suppresses. As a matter of fact, the coupling mechanism enhances the unstable regions as the wave exchanges the energy, and we can deduce that in case of generalized Lorentzian plasma, the coupling between two modes becomes weak to some extent. Moreover for non-zero streaming velocity of ions, the unstable regions tend to grow. After simplifying Eq. (30), we get the mixed shear Alfvén-acoustic mode, i.e.,

$$\omega^2 = \Omega^2_{(dlh)e} + k_z^2 V_{Ae}^2 \left[1 + \frac{\epsilon_\perp \Lambda_i^{-1} \rho_e^2}{c^2 c_\kappa}\right], \tag{31}$$

where $\rho_e^2 = T_i/m_e\Omega_{ce}^2$, and $V_{Ae} = c\Omega_{ce}/\omega_{pe}$. In the limit $\beta_i \ll 1$ and for $k_\perp^2\rho_e^2 \ll 1$, the two modes decouple and we get,

$$\omega^2 = \epsilon_z \frac{\Omega^2_{(dlh)e}}{\omega^2_{pd}}, \tag{32}$$

where $\epsilon_z = \omega_{pd}^2 + k_z^2c^2$ and $\Omega^2_{(dlh)e} = \omega_{pd}^2/F_e$ is the dust lower hybrid frequency which arises due to the hybrid motion of magnetized electrons and unmagnetized dust grains and is referred as a cutoff frequency which gives rise to a limit for the propagation of electromagnetic waves in the presence of dust grains. For graphical representation, we have chosen parameters typical to space dusty environment, for example, we consider $n_{i0} = 10 - 10^4 cm^{-3}$, $n_{d0} = 10 - 10^{-2} cm^{-3}$, $Z_{d0} = 10 - 10^4$, $m_d = 10^5 - 10^8 m_i$. For computational convenience, we introduce dimensionless parameters which are as follows: $\omega = \Omega_c\tilde{\omega}$, $k_z = \Omega_c\tilde{k}_z/V_A$, $V_0 = V_A\tilde{V}_0$. It has been observed that the growth rates of KAW instability are significantly affected by the presence of superthermal population, i.e., instability suppresses due to energetic particles possessed by kappa distribution when compared to its Maxwellian counterpart as shown in **Figure 1**. Similarly, the effect of streaming velocity, dust number density and charge on the growth rates is depicted in **Figures 2–4** respectively. The free energy is associated with the drift motion of ions along the field direction which is responsible for the excitation of KAW. In a streaming plasma the velocity of ions is directly coupled to dust-acoustic waves and through this coupling the maximum growth rate is obtained when the wave exchanges energy through the streaming of ions. Moreover, the presence of dust particles has a noticeable effect on the wave dynamics through dust charge Z_d and number density n_d, i.e., it modifies the wave propagation and excitation. We can observe that Z_d and n_d enhances the growth rates of KAW due to the reason that when dust concentration in plasma is introduced they attach the plasma electrons toward them and the electron loss rate increases which in particular enhances the drift velocity to facilitate the unstable wave structure.

3.5. Dust kinetic Alfvén waves (DKAWs)

DKAWs arise when the dispersion relation of ordinary Alfvén waves is modified by the finite Larmor radius effect of dust. This process is dominated by the collective dynamics of magnetized dust particles. We have investigated shear Alfvén waves and their coupling with dust-acoustic wave by considering magnetized dust and Lorentzian electrons and ions.

The perturbed current and number densities of cold and magnetized dust are obtained by using Eqs. (5) and (6)

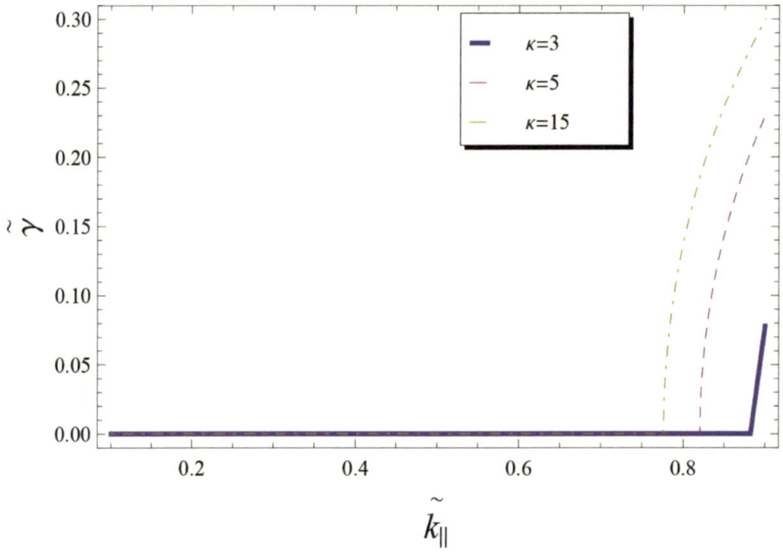

Figure 1. Effect of κ on the imaginary part ($\tilde{\gamma} = \gamma/\Omega_{ce}$) of the dispersion relation.

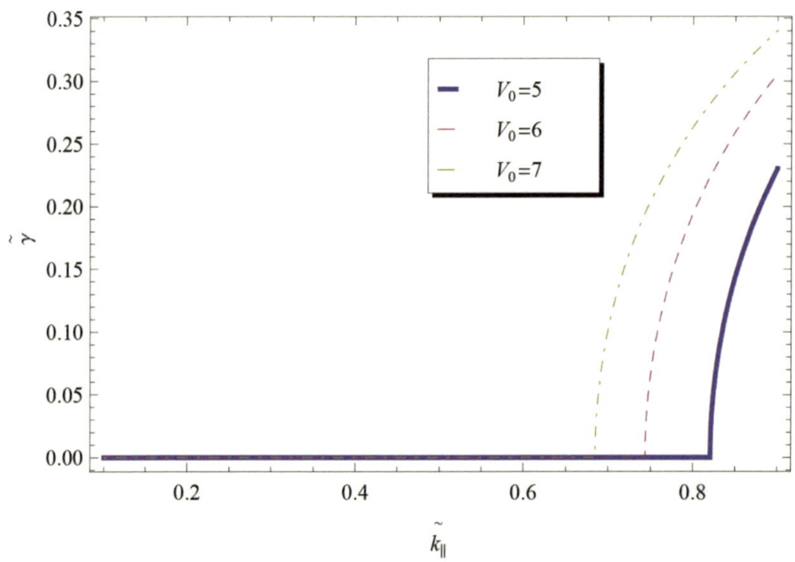

Figure 2. Effect of V_0 on the imaginary part of dispersion relation.

$$n_{d1} = -\frac{n_{d0}Q_{d0}}{m_d}\left[\frac{1}{\omega^2 - \omega_{cd}^2}k_\perp^2\varphi + \left(k_z^2\psi\right)/\omega^2\right] \tag{33}$$

The parallel component of perturbed dust current density turns out to be from Eq. (14) $J_{d1z} = \left(\omega_{pd}^2/\omega\right)\epsilon_0 k_z^2\psi$ and the dispersion relation of kinetic Alfvén wave in the presence of magnetized dust is given by

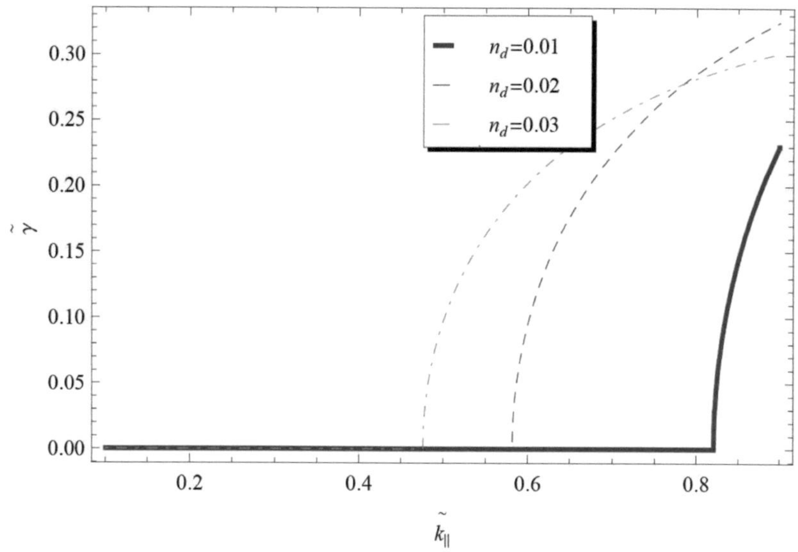

Figure 3. Role of dust number density n_{d0} on the growth rates.

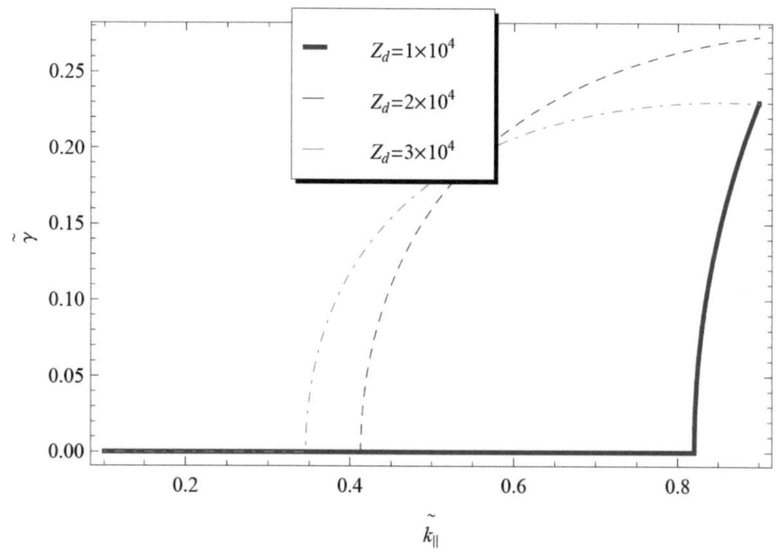

Figure 4. Role of dust charge Z_{d0} on the growth rates.

$$\omega^2 = k_z^2 V_{DA}^2 \left[1 + \frac{\epsilon_\perp \rho_d^2}{c^2 c_\kappa} \right] \tag{34}$$

In the limit $\kappa \to \infty$, we obtain classical results in a Maxwellian plasma.

3.5.1. Lorentzian-type charging currents

The charging equation containing Lorentzian electron and ion currents is

$$\frac{\partial Q_{d1}}{\partial t} = \sum I_{e1}^{\kappa} + I_{i1}^{\kappa} \tag{35}$$

where the electron and ion currents are calculated using a surface integral through the dust grain surface of radius r_d having potential φ_d are given as

$$I_{e1}^{\kappa} = -\tau_1\left[\xi_{e0}(\kappa' + \xi_{e0}Z(\xi_{e0}))m_e\theta_{e\|} + 2e\varphi_d Z_\kappa(\xi_{e0})\right], \tag{36}$$

and

$$I_{i1}^{\kappa} = -\tau_2\left[\xi_{i0}(\kappa' + \xi_{i0}Z(\xi_{i0}))m_i\theta_{i\|} - 2e\varphi_d Z_\kappa(\xi_{i0})\right], \tag{37}$$

where $\tau_1 = (a_d^2\psi)/\left(4m_i\theta_{i\|}(\lambda_{Di}^{\kappa})^2\kappa'\right)$, $\tau_2 = (a_d^2\psi)/\left(4m_i\theta_{i\|}(\lambda_{Di}^{\kappa})^2\kappa'\right)$ and $\lambda_{e.i}^{\kappa} = \left(\frac{1}{c_\kappa}\frac{Te, i}{4\pi n_{e,i0}e^2}\right)^{\frac{1}{2}}$ is the Debye wavelength in superthermal plasma which is much smaller than found for a Maxwellian plasma and has been shown by [39, 41] and a_d is the radius of dust grain.

The Lorentzian charging currents are derived by using Vlasov-kinetic model whose fluid version by Rubab and Murtaza [41] and in the limit $\kappa \to \infty$, our results matched with Das et al., [32]. Now, by putting the value of perturbed dust grain charge, $Q_{d1} = \pm\frac{i}{\omega}\Omega\psi$, in Eq. (1), the dispersion equation of DKAW becomes,

$$\omega^2 = k_z^2 V_{DA}^2\left[1 + \frac{\epsilon_\perp \rho_d^2}{c^2 c_\kappa} \pm i\pi n_{d0}\frac{a_d^2}{k_\|}\right], \tag{38}$$

which clearly shows that charge fluctuation effects are insensitive to the form of the distribution function.

3.5.2. *Modified dust-acoustic wave*

In the limit $\omega^2/k_z^2 V_A^2 \to 0$, the Eq. (16) after simplification turns out to be

$$\omega^2 = \frac{c_\kappa^{-1}k_z^2 C_D^2}{\left(c_\kappa^{-1} + k_\perp^2\rho_d^2 + \pm i\pi n_{d0}\frac{r_d^2}{k_z}\right)}, \tag{39}$$

where $\rho_d^2 = C_D^2/\omega_{cd}^2$, $C_D = (T_{eff}/m_d)^{\frac{1}{2}}$ and $T_{eff} = n_{d0}Z_d^2(T_e n_{i0} + T_i n_{e0})/n_{i0}n_{eo}$. Eq. (39) is the dispersion relation of dust-acoustic wave in a magnetized plasma whose Maxwellian version without dust charge fluctuation effects is given by Mahmood and Saleem [42]. It could be seen that the component of dust velocity in the direction of magnetic field (V_{dz}), which finally turns out to be dust gyroradius, is responsible for the coupling of Lorentzian type DKAW with DAW. When the dust-acoustic wave frequency is very large compared to the dust gyroradius, then the dust is considered to be unmagnetized. In an unmagnetized plasma ($B_0 = 0$) with $T_d = 0$, we get the dispersion relation of Lorentzian dust-acoustic wave (without dust charge fluctuation effects) which is exactly equal as discussed by [40]. The effect of Lorentzian index when growth rates are plotted as function of parallel and perpendicular wave number are depicted through graphical representation in **Figure 5** and **Figure 6** and shows that Maxwellian distribution functions are supportive to enhance the wave frequency.

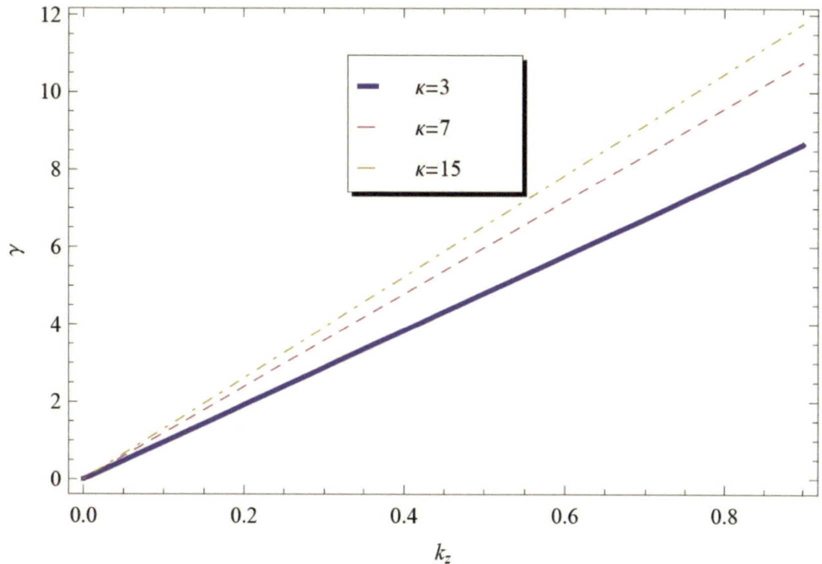

Figure 5. Growth rates ($\tilde{\gamma} = \gamma/\Omega_{cd}$) as a function of k_z for different values of $\kappa = 3, 5, 7$.

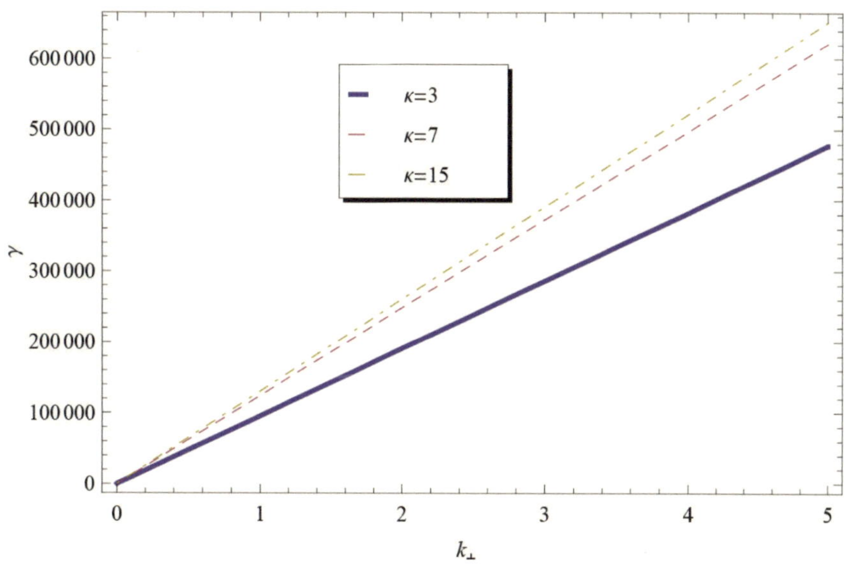

Figure 6. Growth rates ($\tilde{\gamma} = \gamma/\Omega_{cd}$) as a function of $k_\perp$ for different values of $\kappa = 3, 5, 7$.

3.5.3. DKAW: Perpendicular streaming

We consider an electromagnetic dust kinetic Alfvén wave streaming instability in a collisionless electron-ion dusty magnetoplasma. The motion of DKAW is followed by considering thermal and magnetized Lorentzian electrons to be Maxwellian and Lorentzian ions drifting across the external magnetic field $(B_0 \| \hat{z})$ with a constant drift velocity $V_0\hat{x}$, i.e., $(V_0 \perp B_0)$. The dust is considered to be cold and magnetized $(\omega \ll \omega_{cd})$ and the charge on the dust grain surface is taken to be constant. The wave vector associated with the electromagnetic wave lies within xz plane

$$\mathbf{B}_0 = (0,0,B_0), \qquad \mathbf{V}_0 = (V_0,0,0), \qquad \mathbf{k} = (k\sin\theta, 0, k\cos\theta)$$

where

$$k_z = k\cos\theta, \qquad k_\perp = k\sin\theta$$

The distribution function of Lorentzian ions where ions are streaming perpendicular to the field direction is given as,

$$f_{i0}^{\kappa} = A_\kappa\left[1 + \frac{1}{\kappa v_{ti\kappa}^2}\left(v_z^2 + (v_\perp - V_0)^2\right)\right]^{-\kappa-1} ; \kappa > 3/2, \tag{40}$$

$$n_{i1} = -\frac{2e\psi n_{i0}}{m_i v_{ti\kappa}^2}[\kappa' + \eta' Z(\eta')]. \tag{41}$$

where $\eta' = (\omega - k_\perp V_0)k_\perp v_{ti\kappa}$. As there are no ions along the field direction due to perpendicular streaming, therefore we may neglect the ions current density $J_{i1z} = 0$. In the limit $\kappa \to \infty$, our results reduce to Maxwellian distribution.

The dispersion relation with the aid of Eq. (15) is obtained by using Eqs. (23), (24), (33) and (41) in Eqs. (1) and (2), i.e.,

$$\begin{aligned} 1 &+ \frac{2\omega_{pi}^2}{k_z^2 v_{ti\kappa}^2}\left[\left(\kappa' + \eta' Z_\kappa\left(\eta'\right)\right)\chi\right] + \frac{2\omega_{pe}^2}{k_z^2 v_{te\kappa}^2}\left[(\kappa' + \xi_{e0}Z_\kappa(\xi_{e0})) + \frac{\omega v_{te\kappa}}{c^2 k_\parallel^3}\xi(\kappa' + \xi_{e0}Z_\kappa(\xi_{e0}))\right] \\ &+ \frac{k_\perp^2}{k_z^2}\chi(1+F_D) - \frac{\omega_{pd}^2}{\omega^2} = 0, \end{aligned} \tag{42}$$

which is the general dispersion relation of kinetic Alfvén waves in the presence of perpendicular streaming ions and cold and magnetized dust. In the above equation, $F_D = \omega_{pd}^2/\omega_{cd}^2$ is responsible for the magnetized dust part.

For parallel propagation and in the limit $\omega_{pd}^2/c^2k_\perp^2 \ll 1$, $F_D \ll 1$, we get dispersion relation of two stream instability (TSI) in an unmagnetized dusty plasma. In a dust free plasma $\left(\omega_{pd}^2 = 0\right)$, we get the classical well know relation of TSI, while in the absence of streaming ions, i.e., $V_0 = 0$, we obtain the dispersion relation of dust kinetic Alfvén waves

$$\omega^2 = \Omega^2_{(dlh)d} + k_z^2 V_{AD}^2 \left[1 + \frac{\epsilon_\perp \rho_d^2}{c^2 c_\kappa} \Lambda_d\right]. \tag{43}$$

where $\Omega^2_{(dlh)d} = \omega_{pi}/F_D$, $\Lambda_d = n_{d0} z_{d0}/n_{e0}$ and $\rho_d = C_D/\omega_{cd}$. In the limit $k_\perp^2 \rho_d^2 \ll 1$, we obtain modified shear Alfvén wave associated with the hybrid dynamics of the ions and magnetized dust through $\Omega^2_{(dlh)d}$ which provides a cut-off for the EM wave propagation, i.e.,

$$\omega^2 = \Omega^2_{(dlh)d}\left(1 + \lambda_i^2 k_z^2\right) \tag{44}$$

where $\lambda_i = c/\omega_{pi}$.

The dispersion relation for the DKAW instability is found to be dependent on the spectral index κ which means Lorentzian plasma is able to support a number of unstable branches. Lorentzian index is found to be more effective in large wave length limit as compared to small wavelength where the tail of unstable region remains independent of κ. When a large number of dust grains are introduced, it will enhance the loss rate of electrons by attachment on a dust grain surface which reduces the wave activity. At the same time the electron loss rate increases the drift velocity which in turns helps to excite the DKAWs and a further increase will help to stabilize the system. Due to particular choice of equations which involves parallel current density, the ions electromagnetic response cant not take part which limits the existence of ions Weibel instability.

By using the same parameters as above, we have plotted the growth rates as the function of propagation vector for different values of kappa. We have seen that the cross-field streaming of superthermal ions inhibit the growth rate of instability as shown in **Figure 7**. Similarly, β_d is found to support the unstable structure and the instability increases with the value of β_d as shown in **Figure 8**.

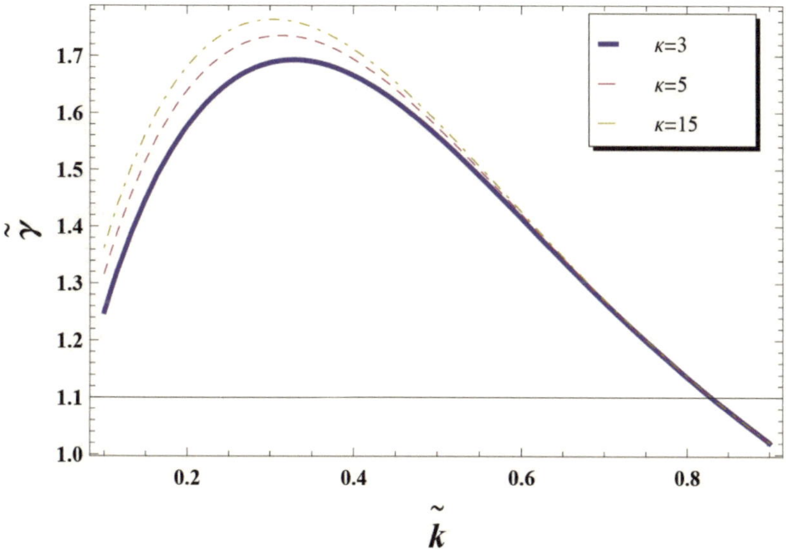

Figure 7. Growth rates ($\tilde{\gamma}$) for perpendicular streaming as a function of wave vector **k** for $\kappa = 3, 5, 15$.

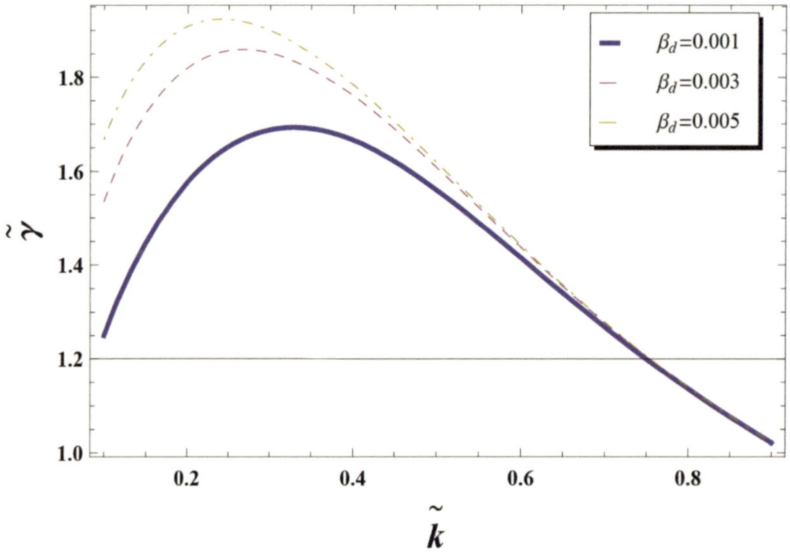

Figure 8. Growth rates ($\tilde{\gamma}$) for perpendicular streaming as a function of wave vector **k** for $\beta_d = 0.001, 0.003,$ and 0.005.

4. Weibel instability in a Lorentzian plasma

The Weibel plasma instability has so many applications in astrophysical [43], and in laboratory plasmas as well [44]. The generation of magnetic field can be explained in the domain of gamma ray burst, galactic cosmic rays and supernovae [45, 48]. For the case of unmagnetized plasma, the Weibel instability [20] has been widely discussed in relativistic and nonrelativistic regimes. In 1989, Yoon [46, 47] generalized his work by using relativistic bi-Maxwellian plasma. Later, Schaerfer [48] have discussed this instability in relativistic regimes of plasma with arbitrary distributions and presented comparison with his previous works which was based on bi-Gaussian distribution functions. The Weibel instability was investigated by Califano [49, 50] with temperature anisotropy, produced by two counterstreaming electron populations. Davidson probed the multi species Weibel instability for the charged beam and intense ions in plasma [51].

In our work, we have derived the analytical expressions and compared the results numerically for the real and imaginary parts of the dielectric constant with the Maxwellian and kappa (κ) distributions under two conditions i.e., $\alpha = \frac{\omega\Theta}{kz} \gg 1$ and $\ll 1$.

By using kinetic model, the linear dispersion relation for Weibel instability in unmagnetized plasma has been derived after solving the linearized, nonrelativistic Vlasov equation as below [52],

$$\omega^2 - c^2k^2 - \omega_{pe}^2 + \pi\omega_{pe}^2\left(\frac{k}{m}\right)\int_{-\infty}^{\infty}\frac{m^3 dv_z}{(\omega - kv_z)}\int_0^{\infty} v_\perp^3 dv_\perp \times \left(\frac{\partial f_{0\kappa}}{\partial v_z}\right) = 0, \tag{45}$$

where f_0 is the distribution function and here we will discuss the different velocity distributions, i.e., Maxwellian distribution and $\kappa-$ distribution functions.

To calculate Weibel instability in a Lorentzian plasma, we use Eq. (21), for zero streaming velocity of particle, i.e., $V_0 = 0$ and using $\left(\frac{\partial f_0}{\partial v_z}\right)$ in Eq. (45), and performing perpendicular integration, we are left with parallel integral which is called modified plasma dispersion function for kappa distribution.

$$\omega^2 - c^2k^2 - \omega_{pe}^2\left(1 - \frac{T_\perp}{T_z}\right) + \frac{\omega_{pe}^2}{\sqrt{\pi}}\left(\frac{T_\perp}{T_z}\right)\left(\frac{\Gamma(\kappa)}{\kappa^{\frac{1}{2}}\Gamma\left(\kappa - \frac{1}{2}\right)}\right)(\alpha)\int_{-\infty}^{\infty}\frac{\left(1 + \frac{x^2}{\kappa}\right)^{-\kappa}}{(x - \alpha)}dx = 0, \tag{46}$$

where $x = \Theta_z^{-1}v_z$, $\alpha = \frac{\omega\Theta_z}{kz}$ and $\Theta_{z,\perp} = \frac{(2\kappa-3)T_{z,\perp}}{\kappa m}$.

Applying same procedure as above and again using Plemelj's formula,

$$\int_{-\infty}^{\infty}\frac{\left(1 + \frac{x^2}{\kappa}\right)^{-\kappa}}{(x - \alpha)}dx = P\int_{-\infty}^{\infty}\frac{\left(1 + \frac{x^2}{\kappa}\right)^{-\kappa}}{(x - \alpha)}dx + i\pi\left(1 + \frac{x^2}{\kappa}\right)^{-\kappa}, \tag{47}$$

the integration of principal part yields

$$P\int_{-\infty}^{\infty}\frac{\left(1 + \frac{x^2}{\kappa}\right)^{-\kappa}}{(x - \alpha)}dx = \frac{\sqrt{\pi}\kappa^{1/2}\Gamma\left(\kappa - \frac{1}{2}\right)}{\Gamma(\kappa)}\left(1 + \frac{\kappa\Gamma\left(\kappa - \frac{3}{2}\right)}{\Gamma\left(\kappa - \frac{1}{2}\right)}\right). \tag{48}$$

The dispersion relation will be solved under two following conditions

For $(\alpha > \mathbf{1})$:

$$\omega^4 - \left(c^2k^2 + \omega_{pe}^2\right)\omega^2 - \omega_{pe}^2\left(\frac{T_\perp}{T_z}\right)\left(v_{tz}^2k_z^2\right) = 0, \tag{49}$$

which shows the real part of Weibel instability is insensitive to the value of Lorentzian index and the imaginary part $i\pi\left(1 + \frac{\alpha^2}{\kappa}\right)^{-\kappa} \to 0$.

For $(\alpha < \mathbf{1})$:

The dispersion relation takes the form

$$\omega^2 - c^2k^2 - \omega_{pe}^2\left(1 - \frac{T_\perp}{Tz}\right) + \frac{\omega_{pe}^2}{\sqrt{\pi}}\left(\frac{T_\perp}{Tz}\right)\left(\frac{\Gamma(\kappa)}{\kappa^{\frac{1}{2}}\Gamma\left(\kappa - \frac{1}{2}\right)}\right)(\alpha)\int_{-\infty}^{\infty}\frac{\left(1 + \frac{x^2}{\kappa}\right)^{-\kappa}}{(x - \alpha)}dx = 0. \tag{50}$$

Now, we define a new plasma dispersion function, i.e.,

$$Z_\kappa^*(\alpha) = \frac{1}{\sqrt{\pi}}\left(\frac{\Gamma(\kappa)}{\kappa^{\frac{1}{2}}\Gamma\left(\kappa - \frac{1}{2}\right)}\right)\int_{-\infty}^{\infty}\frac{\left(1 + \frac{x^2}{\kappa}\right)^{-\kappa}}{(x - \alpha)}dx, \tag{51}$$

the corresponding dispersion relation can be expressed as

$$\omega^2 - c^2k^2 - \omega_{pe}^2\left(1 - \frac{T_\perp}{T_z}\right) + \omega_{pe}^2\left(\frac{T_\perp}{T_z}\right)(\alpha)Z_\kappa^*(\alpha) = 0. \tag{52}$$

We can Solve $Z_\kappa^*(\alpha)$ by taking κ an integer and assuming $\alpha < 1$. So for $\kappa = 3, 5, 7$ we get the following three $Z-$ functions respectively.

$$\begin{aligned} Z_3^*(\alpha) &= \alpha(-1.66 - 0.370\alpha^2 - ....) + \iota(1.539 - 1.539\alpha^2 + ...) \\ Z_5^*(\alpha) &= \alpha(-1.8 - 0.48\alpha^2 - ....) + \iota(1.635 - 1.635\alpha^2 + ...) \\ Z_7^*(\alpha) &= \alpha(-1.98 - 0.59\alpha^2 - ....) + \iota(1.732 - 1.736\alpha^2 + ...) \end{aligned} \tag{53}$$

So the three dispersion relations for the above three corresponding Z-functions are.

For $\kappa = 3,$ we get

$$\begin{aligned} &c^2k^2 + \omega_{pe}^2 - \left(\frac{T_\perp}{T_z}\right)\omega_{pe}^2(1 + 1.66\alpha^2) - i\omega_{pe}^2\left(\frac{T_\perp}{T_\parallel}\right)1.539\alpha = 0 \\ &\gamma = \mathrm{Im}\omega = -(0.649)\frac{k_z v_{Tz}}{\omega_{pe}^2}\left(\frac{T_z}{T_\perp}\right)\left(c^2k^2 + \omega_{pe}^2\left(1 - \frac{T_\perp}{T_z}\right)\right) \end{aligned} \tag{54}$$

Similarly, for $\kappa = 5$ and 7 we obtain the followings

$$\gamma = \mathrm{Im}\omega = -(0.7324)\frac{k_z v_{Tz}}{\omega_{pe}^2}\left(\frac{T_z}{T_\perp}\right)\left(c^2k^2 + \omega_{pe}^2\left(1 - \frac{T_\perp}{T_z}\right)\right) \tag{55}$$

and

$$\gamma = \mathrm{Im}\omega = -(0.8152)\frac{k_z v_{Tz}}{\omega_{pe}^2}\left(\frac{T_z}{T_\perp}\right)\left(c^2k^2 + \omega_{pe}^2\left(1 - \frac{T_\perp}{T_z}\right)\right) \tag{56}$$

Using the Vlasov model, we have derived new dispersion relations based on $\kappa-$ distribution function in an unmagnetized plasma. The analytical expressions for the dielectric constant have been obtained under two conditions i.e., $(\alpha \gg 1)$ and $(\alpha \ll 1)$, which finally give real and imaginary parts respectively. The real part if found to be insensitive to the value of Lorentzian index while imaginary part shows strong dependence on κ. A graphical representation has also been added for the comparison of non-Maxwellian distributions with that of the Maxwellian. The imaginary parts of the dispersion relation obtained above have been plotted for different values of $\kappa-$ showing the variation of the normalized frequencies, i.e., $\frac{\mathrm{Im}\,\omega}{\omega_{pe}}$ against $\frac{ck}{\omega_{pe}}$. Figure exhibits the comparison of the result of kappa distribution with that of the Maxwellian. For small $\kappa,$ the growth rate also reduces but on other hand on increasing the κ value, the growth rate enhances and finally approaches the Maxwellian results which is shown in **Figure 9**.

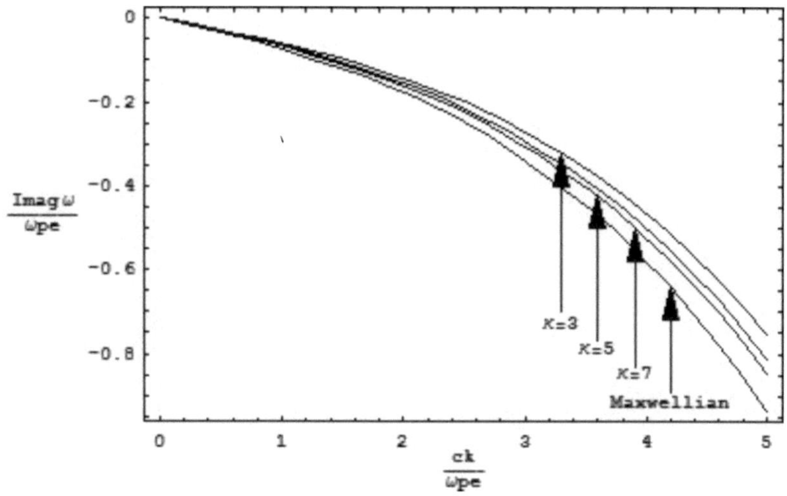

Figure 9. Growth rates of Weibel instability for $\kappa = 3, 5, 7$ and the comparison of results with Maxwellian.

5. Collisional Weibel instability with non-zero magnetic field

The dispersion relation of Weibel instability for transverse waves propagating parallel to magnetic field is obtained as

$$\omega^2 - c^2k^2 - \omega_{pe}^2\left(1 - \frac{\Theta_\perp^2}{\Theta_z^2}\right) + \left(\frac{\omega_{pe}^2}{k\Theta_z}\right)\left(\omega - \left(1 - \frac{\Theta_\perp^2}{\Theta_z^2}\right)(\omega \pm \Omega)\right)Z_\kappa^*(\alpha) = 0 \tag{57}$$

where $\Theta_{z,\perp} = \left(\frac{2\kappa-3}{\kappa}\right)\frac{T_{z,\perp}}{m}$ taking the limit $\alpha > 1$, we obtain

$$\omega^2 - c^2k^2 - \omega_{pe}^2\frac{\omega}{(\omega \pm \Omega)} + \omega_{pe}^2\left(1 - \frac{T_\perp^2}{T_z^2}\right)\left(\frac{k^2v_{T_z}^2}{(\omega \pm \Omega)^2}\right) = 0 \tag{58}$$

We notice that the final expression becomes independent of the spectral index κ.

However, for α small, the dispersion function $Z_\kappa^*(\alpha)$ is obtained by choosing specific values of κ.

For $\kappa = 5, 7$ we get

$$\begin{aligned} Z_5^*(\alpha) &= \alpha(-1.86 - 0.560\alpha^2 - \ldots\ldots) + \iota(1.73 - 1.79\alpha^2 + \ldots..) \\ Z_7^*(\alpha) &= \alpha(-2.05 - 0.734\alpha^2 - \ldots\ldots..) + \iota(1.9 - 1.92\alpha^2 + \ldots\ldots\ldots..) \end{aligned} \tag{59}$$

The imaginary ω therefore becomes

$$\omega = -i(0.79)\left(\frac{kv_{Tz}}{\omega_{pe}^2}\right)\left(\frac{T_z}{T_\perp}\right)\left[c^2k^2 - \omega_{pe}^2\left(1 - \frac{T_\perp}{T_z}\right)\right] \mp \left(1 - \frac{T_z}{T_\perp}\right)\Omega \tag{60}$$

and

$$\omega = -i(0.88)\left(\frac{kv_{Tz}}{\omega_{pe}^2}\right)\left(\frac{T_z}{T_\perp}\right)\left[c^2k^2 - \omega_{pe}^2\left(1 - \frac{T_\perp}{T_z}\right)\right] \mp \left(1 - \frac{T_z}{T_\perp}\right)\Omega \tag{61}$$

for $\kappa = 5$ and 7 respectively.

Considering

$$\omega = \omega_r + i(\omega_i + \nu_e)$$

$$\omega = -i(0.79)\left(\frac{kv_{Tz}}{\omega_{pe}^2}\right)\left(\frac{T_z}{T_\perp}\right)\left[c^2k^2 - \omega_{pe}^2\left(1 - \frac{T_\perp}{T_z}\right)\right] \mp \left(1 - \frac{T_z}{T_\perp}\right)\Omega - \nu_e \tag{62}$$

and

$$\omega = -i(0.79)\left(\frac{kv_{Tz}}{\omega_{pe}^2}\right)\left(\frac{T_z}{T_\perp}\right)\left[c^2k^2 - \omega_{pe}^2\left(1 - \frac{T_\perp}{T_z}\right)\right] \mp \left(1 - \frac{T_z}{T_\perp}\right)\Omega - \nu_e \tag{63}$$

where

$$\nu_e = -\frac{1}{p}\frac{dp}{dt} = -\frac{1}{p}\int_0^\infty f_\kappa v_\perp m_e v_z d\mathbf{v} \tag{64}$$

It is obvious from the above relation that collision frequency for particles obeying kappa distribution differs from that of Maxwellian distribution and is dependent on the value of specie of choice $j = e, i$. It is seen that collision frequency increases with $j = e, i$ and is less for kappa distributed particles than that of the Maxwellian particles. It is therefore justified to use appropriate collision frequency for such Kappa distributed particles.

6. Conclusion

In this chapter, we have described the electromagnetic waves and instabilities in a generalized Lorentzian plasma including particle streaming and finite and anisotropic thermal spread. It allows to grasp the practical understanding of a complex collisionless system in terms of spectra, bulk relative motion and instabilities. In particular, we have focused on kinetic Alfvén waves and instabilities in a dusty and Lorentzian plasma and several types of modes have been identified under various conditions. We have reviewed the kinetic waves and Weibel instabilities in a non-Maxwellian space and astrophysical plasmas by incorporating some basic concepts of dusty environments. We have found that dispersion characteristics involving kinetic Alfvén waves become significantly modified by superthermality effects and dust plasma parameters. The coupling of magnetized dust to the waves due to cyclotron resonance is shown to play a vital role on the wave dynamics. Moreover, the dust grain charging yield some additional plasma currents, which depends on the streaming velocity, Lorentzian index and plasma beta. The Lorentzian index is found to either enhance or quench the electromagnetic instabilities. The dust component is found to play an essential role in wave dynamics, i.e., introducing dust lower hybrid frequency when mobile dust particles are included in the plasma. We have seen that the temperature anisotropy in the distribution function has no effect on the wave characteristics, i.e., the employed model inhibits the temperature anisotropy, but supports the velocity anisotropy. Moreover, a brief analysis on Weibel instabilities in a non-Maxwellian plasma in is also presented.

Kinetic Alfvén turbulence are always present in the streaming solar wind near 1 AU and in situ measurements have confirmed the presence of non-Maxwellian proton distribution function. The present investigations show that the Lorentzian charged particle distributions in space lead to a essentially new physical situation as compared to the plasma with equilibrium distribution functions. Our results of the present analysis opens a new window of investigation to study various streaming and anisotropic modes in different plasma scenarios when Lorentzian distribution function is employed.

Author details

Nazish Rubab[1]* and Sadia Zaheer[2]

*Address all correspondence to: drnrubab@gmail.com

1 Department of Space Science, Institute of Space Technology (IST), Islamabad, Pakistan

2 Department of Physics, Forman Christian College, Lahore, Pakistan

References

[1] Chian AC-L, Rempel EL, Aulanier G, Schmieder B, Shadden SC, Welsch BT, Yeates AR. Detection of coherent structures in photospheric turbulent flows. The Astrophysical Journal. 2014;**786**(1):51

[2] Pierrard V, Lemaire J. Lorentzian ion exosphere model. Journal of Geophysical Research: Space Physics. 1996;**101**(A4):7923-7934

[3] Lazar M, Schlickeiser R, Poedts S, Tautz RC. Counterstreaming magnetized plasmas with kappa distributions–I. Parallel wave propagation. Monthly Notices of the Royal Astronomical Society. 2008;**390**(1):168-174

[4] Lazar M, Tautz RC, Schlickeise R, Poedts S. Counterstreaming magnetized plasmas with kappa distributions–II. Perpendicular wave propagation. Monthly Notices of the Royal Astronomical Society. 2009;**401**(1):362-370

[5] Lazar M, Schlickeiser R, Poedts S. Is the Weibel instability enhanced by the suprathermal populations or not? Physics of Plasmas. 2010;**17**(6):62112

[6] Schlickeise R, Lazar M, Skoda T. Spontaneously growing, weakly propagating, transverse fluctuations in anisotropic magnetized thermal plasmas. Physics of Plasmas. 2011;**18**(1):12103

[7] Rubab N, Erkaev NV, Biernat HK. Dust kinetic Alfvén and acoustic waves in a Lorentzian plasma. Physics of Plasmas. 2009;**16**(10):103704

[8] Rubab N, Erkaev NV, Langmayr D, Biernat HK. Kinetic Alfvén wave instability in a Lorentzian dusty magnetoplasma. Physics of Plasmas. 2010;**17**(10):103704

[9] Rubab N, Erkaev V, Biernat HK, Langmayr D. Kinetic Alfvén wave instability in a Lorentzian dusty plasma: Non-resonant particle approach. Physics of Plasmas. 2011;**18**(7): 73701

[10] Jatenco-Pereira V, CL Chian A, Rubab N. Alfvén waves in space and astrophysical dusty plasmas. Nonlinear Processes in Geophysics. 2014;**21**:405-416

[11] Lazar M, Poedts S, Schlickeiser R, Ibscher D. The electron firehose and ordinary-mode instabilities in space plasmas. Solar Physics. 2014;**289**(1):369-378

[12] Marsch E. Kinetic physics of the solar corona and solar wind. Living Reviews in Solar Physics. 2006;**3**(1):1

[13] Gary SP, Scine EE, Phillips JL, WC. Feldman. JGR. 1994;**99**:391-399

[14] Lakhina GS. Electromagnetic lower hybrid instability driven by solar wind heat flux. Astrophysics and Space Science. 1979;**63**(2):511-516

[15] Christon SP, Mitchell DG, Williams DJ, Frank LA, Huang CY, Eastman TE. Energy spectra of plasma sheet ions and electrons from ~50 eV/e to ~1 MeV during plasma temperature transitions. Journal of Geophysical Research: Space Physics. 1988;**93**(A4):2562-2572

[16] Maksimovic M, Pierrard V, Lemaire JF. A kinetic model of the solar wind with kappa distribution functions in the corona. Astronomy and Astrophysics. 1997;**324**:725-734

[17] Pierrard V, Lamy H, Lemaire J. Exospheric distributions of minor ions in the solar wind. Journal of Geophysical Research: Space Physics. 2004;**109**(A2)

[18] Mourou GA, Tajima T, Bulanov SV. Optics in the relativistic regime. Reviews of Modern Physics. 2006;**78**(2):309

[19] Beskin VS, Gurevich AV, Istomin YAN. Physics of the Pulsar Magnetosphere. Cambridge: Cambridge University Press; 1993

[20] Hasegawa A, Chen L. Kinetic processes in plasma heating by resonant mode conversion of Alfvén wave. The Physics of Fluids. 1976;**19**(12):1924-1934

[21] Weibel ES. Spontaneously growing transverse waves in a plasma due to an anisotropic velocity distribution. Physical Review Letters. 1959;**2**(3):83

[22] Fried BD. Mechanism for instability of transverse plasma waves. Physics of Fluids (1958-1988). 1959;**2**(3):337

[23] Schaefer-Rolffs U, Schlickeiser R. Covariant kinetic dispersion theory of linear waves in anisotropic plasmas. II. Comparison of covariant and noncovariant growth rates of the nonrelativistic Weibel instability. Physics of Plasmas. 2005;**12**(2):22104

[24] Cramer NF. The Physics of AlfvéN Waves. John Wiley & Sons; 2011

[25] Wygant JR, Keiling A, Cattell CA, Lysak RL, Temerin M, Mozer FS, et al. Evidence for kinetic Alfvén waves and parallel electron energization at 4–6 RE altitudes in the plasma sheet boundary layer. Journal of Geophysical Research: Space Physics. 2002;**107**(A8)

[26] Keiling A, Wygant JR, Cattell C, Johnson M, Temerin M, Mozer FS, et al. Properties of large electric fields in the plasma sheet at 4–7 RE measured with polar. Journal of Geophysical Research: Space Physics. 2001;**106**(A4):5779-5798

[27] Keiling A, Wygant JR, Cattell CA, Mozer FS, Russell CT. The global morphology of wave Poynting flux: Powering the aurora. Science. 2003;**299**(5605):383-386

[28] Keiling A, Parks GK, Wygant JR, Dombeck J, Mozer FS, Russell CT, Streltsov AV, LotkoW. Some properties of Alfvén waves: Observations in the tail lobes and the plasma sheet boundary layer. Journal of Geophysical Research: Space Physics. 2005;**110**(A10)

[29] Yukhimuk A, Fedun V, Sirenko O, Voitenko Y. Excitation of fast and slow magnetosonic waves by kinetic Alfvén waves. AIP Conference Proceedings. 2000;**537**:311-316

[30] Nakano T, Nishi R, Umebayashi T. Mechanism of magnetic flux loss in molecular clouds. The Astrophysical Journal. 2002;**573**(1):199

[31] Voitenko Y, Goossens M, Sirenko O, Chian A-L. Nonlinear excitation of kinetic Alfvén waves and whistler waves by electron beam-driven Langmuir waves in the solar corona. Astronomy & Astrophysics. 2003;**409**(1):331-345

[32] Rudakov L, Mithaiwala M, Ganguli G, Crabtree C. Linear and nonlinear landau resonance of kinetic Alfvén waves: Consequences for electron distribution and wave spectrum in the solar wind. Physics of Plasmas. 2011;**18**(1):12307

[33] Das AC, Misra AK, Goswami KS. Kinetic Alfvén wave in three-component dusty plasmas. Physical Review E. 1996;**53**(4):4051

[34] Zubia K, Rubab N, Shah HA, Salimullah M, Murtaza G. Kinetic Alfvén waves in a homogeneous dusty magnetoplasma with dust charge fluctuation effects. Physics of Plasmas. 2007;**14**(3):32105

[35] Treumann RA. Kinetic theoretical foundation of Lorentzian statistical mechanics. Physica Scripta. 1999;**59**(1):19

[36] Summers D, Thorne RM. The modified plasma dispersion function. Physics of Fluids B: Plasma Physics. 1991;**3**(8):1835-1847

[37] Rankin R, Watt CEJ, Samson JC. Self-consistent wave-particle interactions in dispersive scale long-period field-line-resonances. Geophysical Research Letters. 2007;**34**(23)

[38] Galeev AA, Sudan RN. Basic Plasma Physics I. Volume I of Handbook of Plasma Physics. 1983:770 p

[39] Bryant DA. Debye length in a kappa-distribution plasma. Journal of Plasma Physics. 1996;**56**(1):87-93

[40] Lee M-J. Landau damping of dust acoustic waves in a Lorentzian plasma. Physics of Plasmas. 2007;**14**(3):32112

[41] Rubab N, Murtaza G. Dust-charge fluctuations with non-maxwellian distribution functions. Physica Scripta. 2006;**73**(2):178

[42] Mahmood S, Saleem H. Nonlinear dust acoustic and dust kinetic Alfvén waves. Physics Letters A. 2005;**338**(3):345-352

[43] Kennel CF, Sagdeev RZ. Collisionless shock waves in high β plasmas: 1. Journal of Geophysical Research. 1967;**72**(13):3303-3326

[44] Gladd NT. The whistler instability at relativistic energies. The Physics of Fluids. 1983;**26**(4): 974-982

[45] Medvedev MV, Loeb A. Generation of magnetic fields in the relativistic shock of gamma-ray burst sources. The Astrophysical Journal. 1999;**526**(2):697

[46] Yoon PH, Davidson RC. Exact analytical model of the classical Weibel instability in a relativistic anisotropic plasma. Physical Review A. 1987;**35**(6):2718

[47] Yoon PH. Electromagnetic Weibel instability in a fully relativistic bi-Maxwellian plasma. Physics of Fluids B: Plasma Physics. 1989;**1**(6):1336-1338

[48] Schaefer-Rolffs U, Lerche I, Schlickeiser R. The relativistic kinetic Weibel instability: General arguments and specific illustrations. Physics of Plasmas. 2006;**13**(1):12,107

[49] Califano F, Attico N, Pegoraro F, Bertin G, Bulanov SV. Fast formation of magnetic islands in a plasma in the presence of counterstreaming electrons. Physical Review Letters. 2001;**86**(23):5293

[50] Califano F, Cecchi T, Chiuderi C. Nonlinear kinetic regime of the Weibel instability in an electron–ion plasma. Physics of Plasmas. 2002;**9**(2):451-457

[51] Startsev EA, Davidson RC. Electromagnetic Weibel instability in intense charged particle beams with large energy anisotropy. Physics of Plasmas. 2003;**10**(12):4829-4836

[52] Davidson RC. Handbook of Plasma Physics. New York: North-Holland; 1983. pp. 552-557

Chapter 4

Kinetic Theory of Creep and Long-Term Strength of Metals

Alexander Lokoshchenko and Leonid Fomin

Additional information is available at the end of the chapter

http://dx.doi.org/10.5772/intechopen.70768

Abstract

This chapter deals with the simulation of the creep process and the effect of long-term strength of metals, notably, in both uniaxial and complex stress states. A description of a creep experiment and the simplest creep models are presented, that is, the theory of steady creep, the theory of ageing, the theory of flow and the theory of hardening. In creep process simulation, a kinetic theory based on the introduction of structural parameters characterising the state of the metal at a given time is widely used. Among such parameters, metal damage in the creep process, work of stresses on creep deformations (energy version) and concentration of an aggressive medium in the metal were studied. The coupled problem of creep and long tensile strength is also considered taking into account the mutual influence of damage accumulation and one-dimensional diffusion of the aggressive medium. The times to fracture are determined both in the presence of an aggressive medium and in the absence of one. A significant contribution of Soviet (Russian), European, American and Japanese scientists to the development of continuum damage mechanics is highlighted.

Keywords: kinetic theory, continuum damage mechanics, creep, long-term strength, metals, scalar and vector damage parameters, fracture criterion, aggressive medium, diffusion

1. Introduction

In the late 1950s of the twentieth century, two outstanding Soviet scientists Kachanov and Rabotnov concluded that the terms of the deformed solid mechanics (stress and strain tensors and displacement vector) used at that time were insufficient to describe the process of creep and delayed fracture of materials and structural elements under creep conditions. They proposed a new approach to study the long-term strength called kinetic. It is based on the use of Kachanov [1] and Rabotnov's [2] damage parameter and the subsequently developed kinetic theory of creep and long-term strength by Rabotnov [3, 4]. The core of this approach is the introduction of a scalar parameter of damage $\omega(t)$ characterising the structural state of the material at an arbitrary time t. The initial state of the material (at $t = 0$) corresponds to $\omega = 0$; when the destruction occurs, the damage $\omega(t^*)$ takes the value of 1. Later, important results in this area were obtained by A.A. Ilyushin, S.A. Shesterikov, O.V. Sosnin and other Russian scientists.

Following the studies by Kachanov and Rabotnov, the mechanics of continuum fracture started to develop in different countries, mainly for the creep processes of metals. In the last 50 years, the continuum damage mechanics (CDM) has been extensively developing. Representatives of the English schools of mechanics F.A. Leckier and D.R. Hayhurst provided a significant contribution to the development of the theory of damage accumulation. Certain success was also achieved in the studies of Polish scientists M. Chrzanowski and W. Tramczynskii. In France, the CDM fundamentals were formulated using thermodynamic considerations (J. Lemaitre). In the beginning of the 1980s of the twentieth century, this section of mechanics was rapidly developing in the USA as a result of the work of many scientists. Since then, this area has been the centre of attention worldwide both with respect to the development of the fundamentals (some theoretical problems have still remained unsolved) and applications.

2. Creep under uniaxial tension

In the mechanics of solids, it is common to differentiate the materials under investigation by their reaction to load. When under an arbitrary loading process, the material immediately returns to its original state after the load is removed; this means that the material exhibits elastic properties. If after unloading there appear residual deformations that depend only on the load values and the order of their application, but do not depend on the loading rates and durations, such medium is called elastoplastic. When these deformations essentially depend on the duration of loading, such media have creep properties or, more generally, rheological properties.

Below, the simplest form of high-temperature mechanical testing—the uniaxial tensile test—is considered. It is assumed that the entire process of loading or deformation occurs under isothermal conditions.

The tensile loading of a cylindrical specimen made of a homogeneous material with force P will be examined. Supposedly, the specimen L and its cross-sectional area d satisfy the condition $L \gg d$. In this case, it can be further assumed that at a certain distance from the ends the section of the specimen with length l is subjected to uniform tensile loading. It is also assumed that the strains are low, the variation of the cross-sectional area is insignificant and the specimen is deformed uniformly, without the formation of local constrictions. The ratio of force P to the cross-sectional area of the specimen is referred to as stress σ, and the relative variation of length l is referred to as strain $\varepsilon = \Delta l / l$, where Δl is the increment of length l.

This chapter considers a case in which a stress $\sigma(t)$ programme is given in the experiment and the dependency $\varepsilon(t)$ is registered. It should be noted that at this stage, only those programmes that do not lead to the appearance of strain heterogeneities (as a result of the shear lines, neck formation, etc.) are examined.

The chapter contains various terms that are quite familiar to the reader. Therefore, it appears excessive to define such characteristics as deformation, strain tensor, deformation rate, stress and stress tensor.

Previously, it was mentioned that the creep of metals is manifested in the development of the deformation process with time, usually at elevated temperatures. Thus, even in the case of the uniaxial stress state, it is necessary to consider the four macroquantities—temperature, stress, time and strain. Creep characteristics determined in the experiments at constant temperatures can also often be used when evaluating the efficiency of structures at varying temperatures. To determine the dependences describing the creep process, it is usually necessary to use the data obtained in the standard uniaxial tension tests. Creep is mostly common with metals and alloys at absolute temperatures T higher than (0.4–0.5)T^* (T^* is the melting point on the absolute scale, i.e., in Kelvin (K)).

In the creep test, a cylindrical specimen with thermocouples attached to it is secured in the clamps of the loading machine and placed in the furnace. The temperature of the specimen is controlled using the thermocouples, and the results are sent to a tracking system. This system ensures heating of the specimen to the required level, and the temperature is then maintained constant with a specific accuracy. After complete heating of the furnace area, a tensile force is applied to the specimen. This force changes with time under a given law (in most cases, this force is constant or a fractional-constant time function). A strain measurement device is used to record the variation of the length of the specimen with time during continuous recording of the deformation diagram. The elongation of the specimen as a result of the creep of the material is accompanied by a decrease in the cross-sectional area and, consequently, the tensile stress increases continuously at a constant load.

In the tests of the materials characterised by high creep strains (of 4–5% order or more), there are used systems where the load is self-compensated so that the stress in the specimen remains constant. When testing a number of creep-resistant alloys, it appears that the creep strains remain relatively small (approximately 1–2%) up to the moment of fracture. In these conditions, the tests can be carried out at a constant load, and it can be assumed that the stresses remain unchanged during the experiment. The creep experiments show that even for the specimens taken from the same blank part (plate or bar) the creep strain values are greatly scattered for the same values of time (by up to 20–30% or more). The scatter is explained by the specific features of the individual specimens.

Figure 1 schematically shows the curves characterising the strain dependence $\varepsilon(t)$ on time t at the different stresses σ.

The conventional $\varepsilon(t)$ curve corresponding to the average stress level ($\sigma = \sigma_2$) has three distinctive sections as follows: Section 1 with the constantly decreasing creep rate (unsteady creep), Section 2 with a constant (minimum) creep rate (steady-state creep) and Section 3 with accelerating creep preceding fracture. At relatively low stresses ($\sigma = \sigma_1$), the $\varepsilon(t)$ curve can have only the nonsteady section. The curves leading to relatively high stresses ($\sigma = \sigma_3$ and $\sigma = \sigma_4$) may not have the first section, and when $\sigma = \sigma_4$, only the third section is present. All these special features are satisfactorily explained by the presence of at least two structural deformation mechanisms (hardening and softening), which are determined by changes of the dislocation

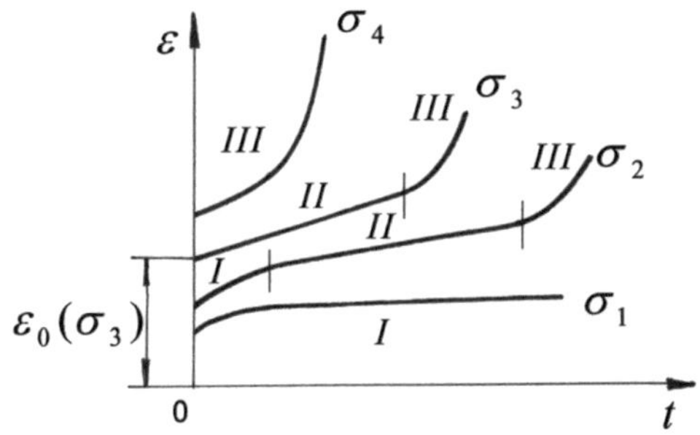

Figure 1. Dependence of strain on time at different stresses.

structure, the vacancy processes, phase transitions, grain size changes in the deformation process and other reasons. The preferential effect of one mechanism in comparison with the others leads to a change of the stages on the creep curve.

When constructing these curves, it is assumed that the loading time of the specimen to a given stress is very short compared to the test time. Therefore, the curves $\varepsilon(t)$ start at the strain corresponding to the 'instantaneous' loading.

The creep theory seeks to determine a relationship between stress σ, time t, creep strain p and temperature T; this relationship, which is universal, should be capable of determining the creep curve $p(t) = \varepsilon(t) - \varepsilon_0(\sigma)$ at the arbitrary laws of stress $\sigma(t)$ and temperature $T(t)$ variations with time.

Different problems of the creep theory have been investigated in a number of monographs ([3–7], and others).

Without loss of generality, here and further on, it is possible to consider isothermal processes occurring at a constant temperature. The transition to other temperatures in creep and long-term strength simulation should be carried out using known temperature-time analogies specified, for example, in [6, 7].

Here, the case is examined where $\sigma(t) = \text{const}$ and the specimen is at the steady-state creep stage $\dot{p}(\sigma, t) = \text{const}$ most of the time. In this case, to describe the behaviour of the material, it is natural to use the relationship of the non-linear viscous flow called the theory of the steady-state creep:

$$\dot{p} = f(\sigma) \tag{1}$$

(the dot above the symbol indicates the differentiation with respect to time t).

The steady-state creep rate $\dot{p}$ is of special importance, because in many technical applications it accounts for the main part of the accumulated creep strain. In most studies, the function $f(\sigma)$ is a power function of the mechanical stress σ:

$$\dot{p} = A\sigma^n, \tag{2}$$

where A and n are the constants of the material.

To describe the first nonsteady section of the creep curve alongside the second one, it is possible to use different theories with the simplest one being the ageing theory:

$$p = F(t, \sigma). \quad (3)$$

The first analytical description of the ageing theory for metals was proposed by E.N. Andrade in 1920:

$$p = At^{1/3} + Bt, \quad (4)$$

where the coefficients A and B depend on stress σ.

An important feature of the creep theory constructed using the relationships explicitly containing the time (this takes place when introducing the hypothesis of ageing) is that they are valid only at constant or relatively slowly changing stresses. For a sudden stress variation, these theories lead to a stepwise variation of the creep strains, too, which is naturally impossible. However, since the ageing hypothesis leads to smaller mathematical difficulties in comparison with the other theories, it is used in calculations with considerable success taking into account the scope of its applicability.

The first attempt to overcome the shortcomings of the above-described ageing theory was considered to be the C.C. Davenport's (1938) hypothesis, according to which relation (3) should be replaced by a relation of the following form:

$$\dot{p} = f(t, \sigma), \quad (5)$$

or taking the elastic properties into account:

$$\dot{\varepsilon} = \frac{\sigma}{E} + f(t, \sigma). \quad (6)$$

This theory is called the theory of ageing in the form of a flow or, briefly, the flow theory.

The simplest consistent assumption used to describe an unstable creep area at a constant temperature is that the creep rate $\dot{p}(t)$ for an arbitrary value of t is determined by the stress σ and the current value of the creep strain:

$$\dot{p} = f(\sigma, p). \quad (7)$$

From Eq. (7), which is the basis of the hardening theory, it follows that the creep rate does not depend explicitly on time t. This fact indicates a significant advantage of the hardening theory over other theories.

The most promising theory in the mechanics of solid media to describe the creep processes of structural metals is the concept of the mechanical equation of state, proposed by Rabotnov [3, 4].

According to this concept, the creep rate $\dot{p}$ of a structurally stable material at every moment of time t depends on the magnitude of applied stress, temperature and the structural state of the material at this moment t. The structural state of the material is characterised by the set of values $q_1, q_2, \dots, q_N$, which are called the structural parameters. The kinetic creep theory consists of the mechanical equation of state

$$\dot{p} = \dot{p}(\sigma, T, q_1, q_2, \dots, q_N) \tag{8}$$

and a system of kinetic equations to determine the parameters q_i. The structural parameters q_i $(i=1, 2, \dots, N)$ used in Eqs. (8) and (3) vary during deformation in accordance with the kinetic equations:

$$dq_i = a_i dp + b_i d\sigma + c_i dt + g_i dT, \tag{9}$$

and the coefficients a_i, b_i, c_i, g_i are the functions of p, σ, t, T and also of $q_1, q_2, \dots q_N$. Relationships (8) and (9) widen the range of theories available to describe greatly varying experimental results. Extensive studies of the creep of metals using the mechanical equation of state in form (8) supplemented by kinetic equations (9) were carried out in a large number of researches by Rabotnov and his colleagues [8].

A creep process with a stepwise increase of the stress $\sigma(t)$ is investigated below. For an analytical description of the creep curve after changing the mechanical stress, the system of equations (8) and (9) is used. Consider the energy version of the kinetic theory.

At $N=1$, $a_1=a_1(\sigma)$, $b_1=0$, $c_1=0$, $g_1=0$, in a partial case $a_1(\sigma)=\sigma$ and

$$dq = \sigma dp \tag{10}$$

Here, the parameter q is the work of stresses acting at the creep strains. Application of this version in the theory to describe the creep process based on the energy approach has been widely used in the studies by Sosnin and his colleagues [9].

In [10], the results of experimental verification of this version of the theory for D16AT aluminium alloy at the temperature of 150°C are presented. The circles in **Figure 2** show the experimental creep curves at $\sigma_1=150$ MPa and $\sigma_2=250$ MPa.

When simulating these experimental data, the simplest form of the hardening theory is initially examined assuming that it is expressed as follows:

$$\dot{p} = p^{-\alpha} f(\sigma), \ \ \alpha > 0. \tag{11}$$

It is further assumed that $f(\sigma)$ has the form of a power function. Therefore,

$$\dot{p} p^{\alpha} = B\sigma^{n}. \tag{12}$$

Integration of differential equation (12) separately over the intervals $0<t<t_1$ and $t>t_1$ results in the following expressions:

$$p(t) = \left[B(\alpha+1)\sigma_1^n\right]^{\frac{1}{\alpha+1}} t^{\frac{1}{\alpha+1}} \text{ if } 0 \leq t < t_1, \tag{13}$$

$$p(t) = \left[B(\alpha+1)\sigma_2^n(t-t_1) + p_1{}^{\alpha+1}\right]^{\frac{1}{\alpha+1}} \text{ if } t > t_1, \tag{14}$$

where $p_1 = p(t_1) = \left[B(\alpha+1)\sigma_1^n\right]^{\frac{1}{\alpha+1}} t_1^{\frac{1}{\alpha+1}}$.

Now, the version of equations (8) and (9) is selected in such a way that at a constant stress ($\sigma_1 = \sigma_2$) they coincide with (12):

$$\dot{p} q^\alpha = B\sigma^{n+\alpha}. \tag{15}$$

In the first creep section under the effect of the stress $\sigma = \sigma_1$, there is obtained $q(t) = \sigma_1 p(t)$. Therefore, differential equations (12) and (15) coincide and, consequently, the creep curve, corresponding to the defining Eq. (15) at $0 < t < t_1$ coincides with the curve (13). After additional step loading,

$$q(t) = \sigma_2 p(t) - p_1(\sigma_2 - \sigma_1). \tag{16}$$

Substituting $q(t)$ (16) into differential equation (15) and integrating this equation at $t > t_1$ taking into account the initial condition $p(t_1) = p_1$, it is obtained that

$$p(t) = \frac{\sigma_2 - \sigma_1}{\sigma_2} + \left[B(\alpha+1)\sigma_2^n(t-t_1) + \left(\frac{\sigma_1 p_1}{\sigma_2}\right)^{\alpha+1}\right]^{\frac{1}{\alpha+1}}. \tag{17}$$

The dashed line in **Figure 2** shows the creep curve calculated using Eq. (14), and the solid line represents the curve calculated using Eq. (17). Evidently, the hardening theory set up by Eq. (15) taking into account kinetic Eq. (10) describes the experimental data more efficiently than the simplest hardening theory (12).

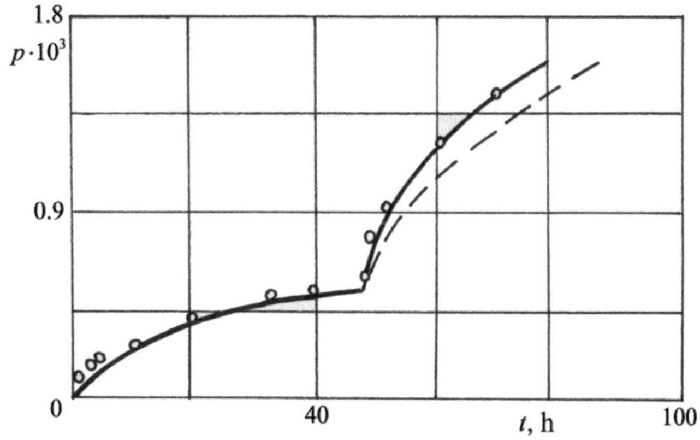

Figure 2. Creep of D16AT alloy [10].

Considering the system of Eqs. (8) and (9) with two kinetic parameters—in addition to the generally accepted hardening measure $q_1=p$, we introduce the second parameter q_2 determined by the following relationship [11]:

$$dq_2 = \begin{cases} pd\sigma & npu \;\; d\sigma > 0, \\ 0 & npu \;\; d\sigma \leq 0. \end{cases} \tag{18}$$

If at the initial moment of time $(t=0, p=0)$, there is applied stress σ_1 that subsequently remains constant, and then $q_2=0$ if $0<t<t_1$.

At a stepwise increase of stress at the time t_1, the parameter q_2, according to Eq. (18), receives the increment $\Delta q_2=p_1(\sigma_2-\sigma_1)>0$. The creep law in [11] is represented by the equation

$$\dot{p}p^{\alpha} = k\exp\left(\frac{\sigma}{A}+\frac{q_2}{B}\right). \tag{19}$$

Eq. (19) implies that additional introduction of the second kinetic parameter q_2 according to (18) leads to a higher creep rate increase after the instantaneous loading compared with the standard hardening theory (12). In case of a step load, it follows from kinetic equation (18) that the parameter q_2 does not change.

Figure 3 shows the experimental points [11] obtained on samples of D16T alloy, tested at the stress of $\sigma_1=80$ MPa for the time $t_1=24$ h and at $\sigma_2=160$ MPa and at $t>t_1$ (temperature of 200°C). The dashed line corresponds to standard hardening theory (12), the solid line—to the theory with two kinetic parameters described by Eq. (19).

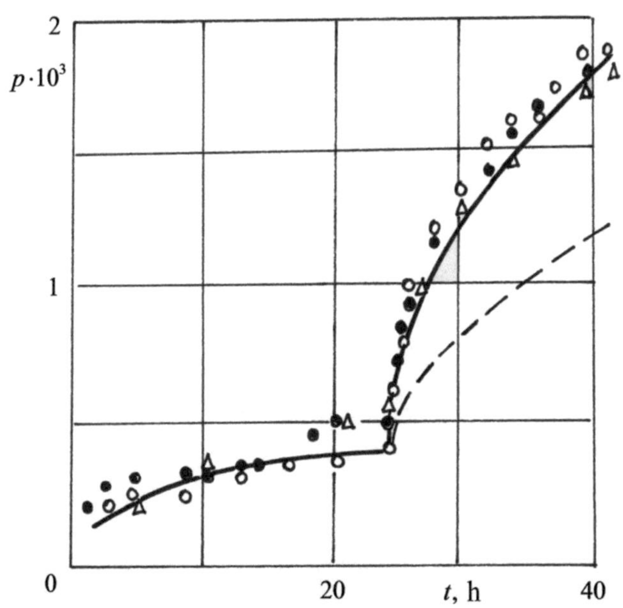

Figure 3. Creep of D16T alloy [11].

These considerations show that the concept of equation of the mechanical state (8) with the system of the kinetic equation (9) proposed by Rabotnov to determine the structural parameters is highly promising for describing different special features of the material behaviours in the creep conditions.

3. Long-term strength under uniaxial tension

In the majority of cases when the levels of temperatures and stresses in the experiments are quite high, the deformation process over time ends with the fracture of the specimen. This moment is characterised by certain time t^* determined by the given values of stress σ and temperature T. If a sufficiently large series of experiments is carried out, then for a number of temperature values T, it is possible to construct a series of curves $t^*(\sigma)$ which are referred to as the long-term strength curves. It should be noted that the actual experimental data for the majority of metals and alloys are greatly scattered.

Typical long-term strength curves are shown schematically in **Figure 4**. In the first case, the points in the logarithmic coordinates are distributed over a single straight line. In the second case, the diagram consists of two straight sections. In this case, section AB of the diagram corresponds to ductile fracture, and section BC corresponds to brittle fracture. The diagram does not always consist of two straight lines characterised by a distinctive intersection point. Sometimes, a curvilinear transition region occurs between the straight lines AB and BC indicated by the dashed line—the region of mixed fracture.

A general approach to the fracture problem implies that the value ω (the extent of cracking) is regarded as a structural parameter. Therefore, the creep process is described by the creep equation

$$\dot{p} = f(\sigma, \omega), \quad p(t = 0) = 0, \tag{20}$$

and the fracture process is described by the kinetic equation of gradual fracture

$$\dot{\omega} = \varphi(\sigma, \omega), \quad \omega(t = 0) = 0. \tag{21}$$

If the fracture is accompanied by a small elongation only, the stress σ at a constant tensile loading force may be regarded as constant ($\sigma(t) = \sigma = \text{const}$). In this case, Eq. (21) is integrated

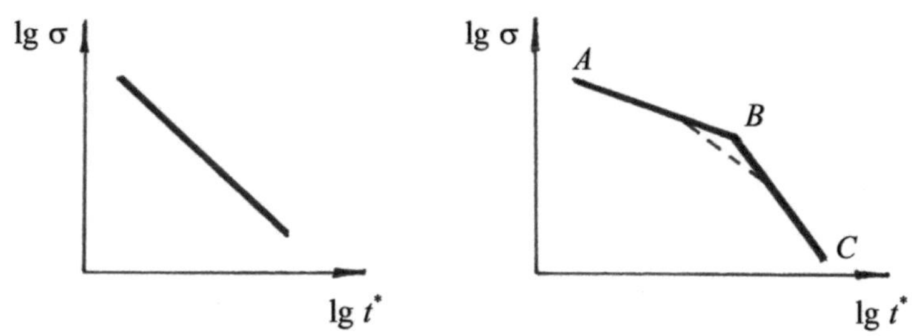

Figure 4. Typical long-term strength curves.

independently of (20) and provides the equation $\omega = \omega(\sigma, t)$. Therefore, the moment of fracture t^* is determined as the value $t = t^*$ at which $\omega = 1$. Further, substitution of $\omega = \omega(t)$ into Eq. (20) leads to the creep curve equation $p(t)$. The following simplest Eqs. (20) and (21) can be used

$$\dot{p} = A\sigma^n(1-\omega)^{-k}, \tag{22}$$

$$\dot{\omega} = B\sigma^m(1-\omega)^{-s}. \tag{23}$$

As previously, here σ is the mean macrostress determined by the tensile force divided by the area of the sample. The deformation process is completed at the moment of fracture $t = t^*$ corresponding to the value of the parameter $\omega(t^*) = 1$. Integration of the system of Eqs. (22) and (23) at $s + 1 - k > 0$ provides the equation of the creep curve

$$p = p^*\left[1 - \left(1 - \frac{t}{t^*}\right)^{\frac{s+1-k}{s+1}}\right], \tag{24}$$

where the time to fracture t^* and the respective limiting creep strain p^* depend on the stress σ and the material constants as follows:

$$t^* = [B(s+1)\sigma^m]^{-1}, \quad p^* = \frac{A}{B(s+1-k)}\sigma^{(n-m)}. \tag{25}$$

Some tests of metals in the creep conditions before fracture showed a non-monotonic change of the limiting creep strain p^* corresponding to the moment of fracture t^* in the investigated range of the constant tensile stress σ. In [12], it was reported that when simulating the non-monotonic dependence $p^*(\sigma)$, various functional relationships should be used to take into account the effect of stress on the creep rate and damage accumulation rate.

To describe the creep process at a constant stress up to the moment of fracture and to determine deformation p^*, consider the exponential dependence of the creep rate on the stress:

$$\dot{p} = C(\mathrm{sh}(\sigma/c))(1-\omega)^{-n} \tag{26}$$

and the power-law dependence of the change rate of the kinetic parameter ω on σ:

$$\dot{\omega} = D\sigma^k(1-\omega)^{-k}, \quad k > 1. \tag{27}$$

Consider the relation $\dot{p}/\dot{\omega}$ and integrate it. As a result, in accordance with Eqs. (26) and (27), the following dependence of the limiting strain p^* on the level of mechanical stress σ is obtained:

$$p^* = \frac{C}{D} \cdot \frac{\mathrm{sh}(\sigma/c)}{\sigma^k}. \tag{28}$$

For relatively small values σ, the dependence $p_i^*(\sigma)$ in (28) is decreasing; for sufficiently large values σ, this dependence is increasing. Consequently, at some intermediate value of the stress, the limiting strain is minimal. The condition

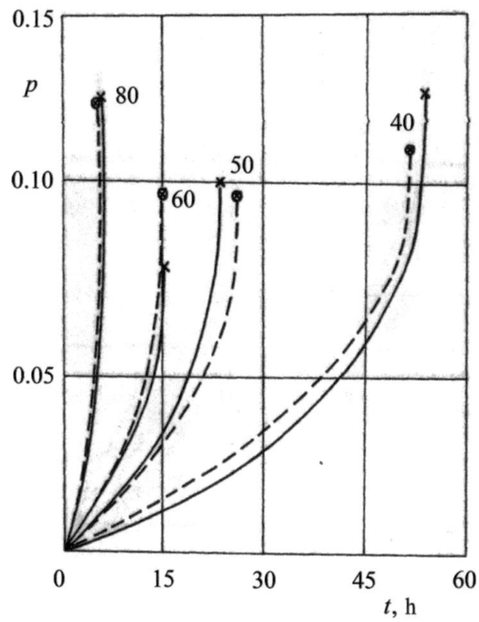

Figure 5. Creep curves of Cr18Ni10Ti steel with the non-monotonic dependence $p^*(\sigma)$.

$$(dp^*/d\sigma)|_{\sigma=\sigma_i} = 0 \tag{29}$$

with due account for Eq. (28) allows determining this value for kinetic equation (27):

$$\text{th}(\sigma/c) = \sigma/(kc). \tag{30}$$

Figure 5 shows the creep curves for stainless steel Cr18Ni10Ti at 850°C and at stresses $\sigma = 40 - 80$ MPa [12]. The experimental curves are shown by solid lines, and the theoretical curves are shown by dashed lines ($n = 2.28$, $k = 3.1$, $C = 2.02 \cdot 10^{-4}$ h^{-1}, $D = 5.1 \cdot 10^{-8}$ (MPa)$^{-3.1}$h^{-1}, $c = 17.8$ MPa).

When describing a series of creep curves before fracture with an internal minimum of the dependence $p^*(\sigma)$, the types of dependencies $\dot{p}$ and $\dot{\omega}$ on σ should be interchanged.

4. Creep under complex stress state

In the previous paragraphs, attention was given to the main assumptions regarding the metal creep phenomenon, and the basic models describing the creep of bars under uniaxial tensile loading were described. In the determination of the mechanical behaviour of the structural elements in the creep conditions, however, it is necessary, as a rule, to consider the multiaxial complex stress state. Experimental studies of creep in these conditions are associated with considerable technical difficulties, and, therefore, the currently available experimental data are not extensive and do not allow a reliable justification of any creep theory in the complex stress state conditions. To determine the characteristics of the material, tests are usually carried out on thin-walled, tubular specimens—the stress state in the specimens is usually generated by a combination of tensile loading with torsion or tensile loading with internal pressure. To generate the quasi-homogeneous stress state in the tubular specimens, the latter must be thin walled. In rare instances, experimental studies of creep in the complex stress state are carried out on a rectangular plate in biaxial loading conditions.

As noted in the beginning of the chapter, for the readers familiar with the basics of the theory of elasticity and plasticity, the concepts of stress tensors, strains and strain rates are also known.

Creep strains in the complex stress state p_{ij}, just as in the case of uniaxial tension, mean the differences between total deformations ε_{ij} and instantaneous deformations arising under quasi-static loading.

The first results of researching the creep of metals under multiaxial stress state conditions were published in the 1930s of the twentieth century. The basic principles of constructing the creep theory are considered in various monographs and textbooks.

Creep theories for the complex stress state usually consider the following three hypotheses:

1. The volume deformation is elastic (the material is always considered as incompressible).
2. The hypothesis of the proportionality of the stress deviators and creep strain rates (the flow-type theory) or the deviators of the stresses and strains (deformation theory).
3. A functional relationship between the second invariants of the tensors of the stresses and creep strain rates (or the stress and strain tensors) is assumed to have such a form that the relationships of one of the well-known creep theories are fulfilled in the partial case of uniaxial tensile loading. The presence of such a relationship implies that the dependence of the intensity of creep strain rates $\dot{p}_u$ on stress intensity σ_u (or intensity of creep strains p_u on σ_u) is the same at different types of the stress state (i.e., the 'single curve' hypothesis is satisfied). Sometimes, the dependence between the stress and strain intensities can be replaced by a similar dependence between the maximum tangential stress and the maximum shear stress.

Different approaches to the construction of metal creep models under the conditions of the multiaxial complex stress state have been discussed in a number of monographs and journal articles (e.g., [3–7]).

As an example of such models, consider the steady-state creep model.

The hypothesis of proportionality of stress deviators and creep strain rates for an incompressible body can be presented as follows:

$$\dot{p} = 0, \quad \dot{p}_{ij} = \frac{3}{2} \cdot \frac{f(\sigma_u)}{\sigma_u} s_{ij}, \quad \begin{matrix} s_{ij} = \sigma_{ij} - \sigma\delta_{ij}, \\ \delta_{ij} = \begin{cases} 1, & i = j, \\ 0, & i \neq j \end{cases} \end{matrix} \quad \sigma = (1/3)\sum_{k=1}^{3} \sigma_{kk}, \quad \dot{p}_u = f(\sigma_u). \tag{31}$$

The relationship of the tensors σ_{ij} and $\dot{p}_{ij}$ in the Cartesian coordinates can have the form:

$$\begin{cases} \dot{p}_{xx} = \frac{3}{2} \cdot \frac{f(\sigma_u)}{\sigma_u} (\sigma_{xx} - \sigma), \dots, \\ \dot{p}_{xy} = \frac{3}{2} \cdot \frac{f(\sigma_u)}{\sigma_u} \sigma_{xy}, \dots . \end{cases} \tag{32}$$

The simplest version of the theory of steady-state creep under uniaxial tensile loading is the power dependence $\dot{p}_1 = B\sigma_1^n$; in a general case, the identical relationship of the second invariant of the corresponding tensors has the form

$$\dot{p}_u = f(\sigma_u) = B\sigma_u^n. \tag{33}$$

Creep equation (31) is the equation of the non-linear viscous flow. They were derived by R. Bailey in 1935 and J. Marin in 1942.

5. Long-term strength under complex stress state

In the kinetic theory, damage accumulation is investigated as a process of gradual fracture of the material. In many studies of Russian and foreign scientists when examining the multiaxial complex stress state, special attention is given to the damage parameters that are not only of the scalar but also of vector and tensor nature. Modern versions of the kinetic theory allow describing the deformation and long-term fracture of metals in nonproportional loading taking into account the anisotropy of metal properties, using the theory to solve technological problems, etc.

Consider the uniaxial ($\sigma_1 = \sigma_0 > 0$, $\sigma_2 = \sigma_3 = 0$) and equiaxed plane ($\sigma_1 = \sigma_2 = \sigma_0 > 0$, $\sigma_3 = 0$) stress states for the same stress level σ_0 (**Figure 6**). The available test results show that the time to

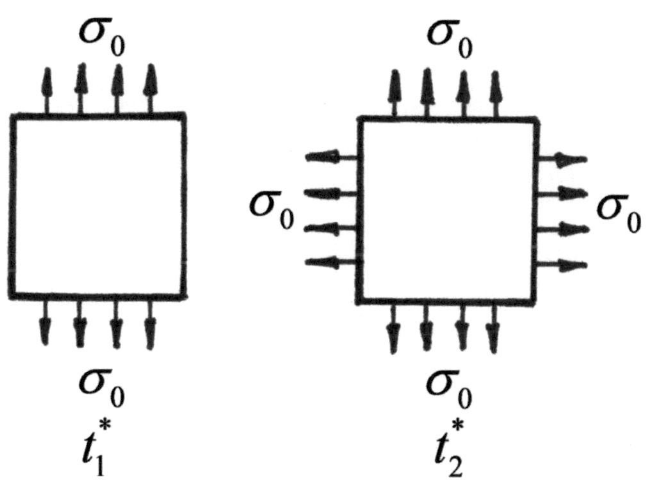

Figure 6. Uniaxial and equiaxed planar stress states.

fracture t_1^* in the uniaxial tensile loading is considerably greater than the time to fracture t_2^* in biaxial loading under these conditions ($s = t_1^*/t_2^* > 1$) [13, 14].

All the investigated experimental data indicate that the addition to the axial tensile loading stress of a transverse tensile stress of the same magnitude decreases the time to fracture several times.

Calculations of the long-term strength of structural members loaded under the conditions of the stationary multiaxial complex stress state are usually carried out using a criterial approach. In this approach, the only characteristic of the stress state taken into account is the so-called equivalent stress σ_e. This characteristic is represented by different combinations of the stress tensor components with a distinctive mechanical meaning such as the maximum tensile stress, the intensity of tangential stresses, the difference of the maximum and minimum main stresses and other expressions. Since these equivalent stresses coincide ($\sigma_e = \sigma_0$) for the investigated uniaxial and biaxial tensile loading, it is not possible to obtain different values of t_1^* and t_2^* using the criteria relationship $t^* = t^*(\sigma_e)$. Further, two versions of the systems of kinetic equations with the vector damage parameter to describe different values of the time to fracture t_1^* and t_2^* are examined.

First, the dependence of the time to fracture t^* on the type of stress state will be described taking into account the instantaneous damage for an isotropic material. This will be carried out using the generalisation of the vector approach taking into account the damage accumulated during loading. One of the possible models for describing different times t_1^* and t_2^* is the system or relationships:

$$d\omega_i = \frac{d\varphi(\sigma_i)}{d\sigma_i} \cdot d\sigma_i + f(\sigma_i) \cdot dt, \quad i = 1, 2. \tag{34}$$

where function $\varphi(\sigma_i)$ characterises the projection of ω_i on the x_i axis of the vector of damage accumulated during loading; $f(\sigma_i)$ is a constant rate of increase of the projection ω_i with time t. In uniaxial tensile loading, from relationships (34), it follows that

$$\omega_1(t) = \varphi(\sigma_0) + f(\sigma_0) \cdot t, \quad \omega_2 = 0, \quad t_1^* = [1 - \varphi(\sigma_0)]/f(\sigma_0), \tag{35}$$

and in the case of the equal biaxial tensile loading, relationships (34) provide

$$\omega_1(t) = \omega_2(t) = \varphi(\sigma_0) + f(\sigma_0) \cdot t, \quad t_2^* = \left[\sqrt{2}/2 - \varphi(\sigma_0)\right]/f(\sigma_0). \tag{36}$$

Relationships (35) and (36) show that the instantaneous value of damage $\varphi(\sigma_0)$ should be in the range $0 < \varphi(\sigma_0) < \sqrt{2}/2$, and the ratio

$$s = t_1^*/t_2^* = (1 - \varphi(\sigma_0))/\left(\sqrt{2}/2 - \varphi(\sigma_0)\right) \tag{37}$$

should exceed $\sqrt{2}$ at any values of σ_0 in this range. As an example of using Eq. (34), the results of tests [13] which at $\sigma_0 = 56.2$ MPa lead to the following values are considered: $t_1^* = 900$ h and

$t_2^* = 280$ h. At $\varphi(\sigma_0) = 0.57$ and $f(\sigma_0) = 4.78 \cdot 10^{-4}$ h^{-1}, the theoretical values of t_1^* and t_2^* calculated from the relationship (36) coincide with the respective experimental values.

The results show that the ratio s depends only on the level of damage $\varphi(\sigma_0)$ accumulated under quasi-static loading. Taking into account the instantaneous damage in the form (5.1) makes it possible to describe the experimental data only for $s \geq \sqrt{2}$. In this case, the result does not depend on the nature of damage accumulation during creep.

Now, the dependence of time to fracture t^* will be described taking into account the anisotropy of the material. When determining the long-term strength of thin-walled pipes, it is taken into account that in the process of such pipes manufacturing the material may acquire anisotropic strength properties (difference in material properties in different directions). To facilitate a quantitative analysis, there are introduced anisotropy coefficients α_1 and α_2 characterising the anisotropy of the instantaneous and long-term strength properties, respectively,

$$f(\sigma_{zz}/\alpha_1) = f(\sigma_{\theta\theta}), \quad \varphi(\sigma_{zz}/\alpha_2) = \varphi(\sigma_{\theta\theta}), \tag{38}$$

Further on, these coefficients will be assumed to be equal to $\alpha_1 = \alpha_2 = \alpha \geq 1$.

In Eq. (38), the components of the stress tensor in a cylindrical coordinate system are used. Anisotropy analysis of the long-term strength characteristics of metals is described in detail in [6, 7].

Consider the kinetic equation for the components of the damage vector ω_i in the following form:

$$d\omega_i = \omega_i^{-1} \cdot \left[d\varphi(\widehat{\sigma}_i) + f_i(\widehat{\sigma}_i)dt\right], \quad i = z, \ \theta \tag{39}$$

$\widehat{\sigma}_i$ represents the transformed main stresses $\widehat{\sigma}_{zz} = \sigma_{zz}/\alpha$, $\widehat{\sigma}_{\theta\theta} = \sigma_{\theta\theta}$ $(\alpha > 1)$. Simple transformations of Eq. (39) provide the relations for the square of the damage vector length under uniaxial and biaxial stretching, respectively:

$$\begin{aligned} \omega^2 &= 2[\varphi(\sigma_0/\alpha) + f(\sigma_0/\alpha) \cdot t], \\ \omega^2 &= 2[\varphi(\sigma_0/\alpha) + \varphi(\sigma_0) + f(\sigma_0/\alpha) \cdot t + f(\sigma_0) \cdot t]. \end{aligned} \tag{40}$$

Eq. (40) allows deriving the following expressions for t_1^*, t_2^* and s:

$$\begin{aligned} t_1^* &= \frac{0.5 - \varphi(\sigma_0/\alpha)}{f(\sigma_0/\alpha)}, \quad t_2^* = \frac{0.5 - \varphi(\sigma_0/\alpha) - \varphi(\sigma_0)}{f(\sigma_0/\alpha) + f(\sigma_0)}, \\ s &= \frac{t_1^*}{t_2^*} = \frac{[f(\sigma_0/\alpha) + f(\sigma_0)]}{f(\sigma_0/\alpha)} \cdot \frac{[0.5 - \varphi(\sigma_0/\alpha)]}{[0.5 - \varphi(\sigma_0/\alpha) - \varphi(\sigma_0)]}. \end{aligned} \tag{41}$$

It is obvious from expression (41) that the value s is greater than 1 for any kinds of functions $f(x)$ and $\varphi(x)$, values $\alpha > 1$ and levels of the stress state σ_0.

6. Creep and long-term strength in the presence of aggressive medium

Forecasting the life of materials and structural elements residing under conditions of prolonged high-temperature loading in the presence of an aggressive medium is an extremely topical task to ensure reliability throughout the working lifespan. In this section, the influence of an aggressive medium on the creep and the long-term strength of materials and structural elements are considered [6, 7, 15–18]. This influence is determined by the diffusion penetration of the medium elements into the material reducing the duration of operability.

Here, an analysis of the long-term tensile strength of a long, thin rod of a rectangular cross section with thickness H_0 and width b, with $H_0 << b$ residing in an aggressive medium, is presented [17]. Following the accepted geometric dimensions, a one-dimensional diffusion process in a rod along the axis x arranged along the thickness H_0 is considered. Together with the process of the aggressive medium diffusion in the rod, the accumulation of damage in the rod material during the creep process is taken into account. Examine the coupled problem of determining the long-term strength of a tension rod under the condition of mass transfer on its surface. This problem statement takes into consideration a mutual dependence of the concentration level of the medium $c(x, t)$ in the rod material and the amount of accumulated damage ω. To this end, the dependence of the diffusion coefficient D on the level of damage ω is considered. For simplicity, the dependence $D(\omega)$ is assumed to be linear:

$$D(\omega) = D_0(1 + k\omega), \quad D_0 = \text{const}, \ k = \text{const}. \tag{42}$$

The following dimensionless variables,

$$\bar{t} = \frac{48D_0}{H_0^2}t, \quad \bar{x} = 2x/H_0, \quad \bar{c} = c/c_0, \quad \bar{A} = \frac{A\sigma_0^n H_0^2}{48D_0}, \tag{43}$$

are introduced, where t is the time, c is the concentration, σ_0 is the nominal stress and A and n are the material constants in the power law of creep.

Consider a simplified problem statement where $\omega(\bar{t})$ is understood as an integral mean damage in the rod cross section. The rod fracture criterion is considered as $\omega(\bar{t}^{**}) = 1$.

In this case, the system of equations in the adopted dimensionless variables consisting of the parabolic diffusion equation and the kinetic equation of damage accumulation has the following form:

$$\begin{cases} \dfrac{\partial \bar{c}}{\partial \bar{t}} = \dfrac{1}{12}(1 + k\omega)\dfrac{\partial^2 \bar{c}}{\partial \bar{x}^2}, \\ \dfrac{d\omega}{dt} = \bar{A}(1 - \omega)^{-n} \cdot f(\bar{c}_m(\bar{t})), \end{cases} \tag{44}$$

The linear form of the function $f(\bar{c}_m(\bar{t})) = 1 + a\bar{c}_m(\bar{t})$ is chosen,

where $\bar{c}_m(\bar{t}) \equiv \int_0^1 c(x, t)dx$ is the dimensionless integral mean concentration of the medium in the rod material.

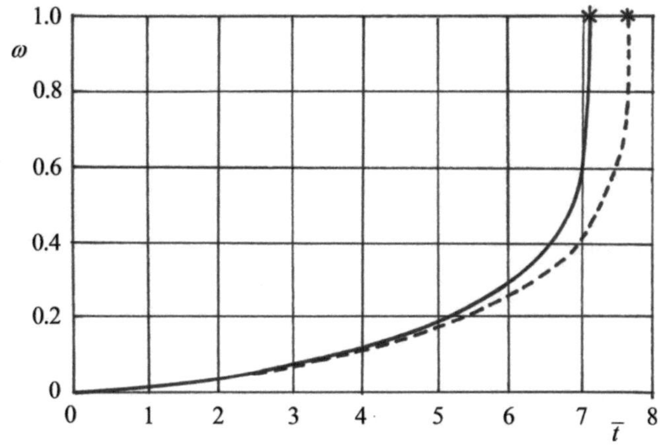

Figure 7. Dependences $\omega(\overline{t})$ for constant and variable diffusion coefficients.

The initial and boundary conditions are assumed to be as follows:

$$\omega(0) = 0,\ \overline{c}(\overline{x}, 0) = 0,\ \frac{\partial \overline{c}}{\partial \overline{x}}(1, \overline{t}) = \overline{\gamma}\left[\overline{c}(1, \overline{t}) - 1\right],\ \frac{\partial \overline{c}}{\partial \overline{x}}(0, \overline{t}) = 0, \tag{45}$$

where $\overline{\gamma}$ is the dimensionless mass transfer coefficient.

The following values of the constants are used in calculation

$$n = 3,\ \overline{\gamma} = 1,\ k = 4,\ \overline{A} = 0.01,\ a = 9.5. \tag{46}$$

The dependence $\omega(\overline{t})$ for constants (46) is shown in **Figure 7** by a solid line, while the time to fracture $\overline{t}^{**} = 7.16$. In **Figure 7**, the dashed line additionally shows the dependence $\omega(t)$ corresponding to the solution of the system of Eq. (44) at $k=0$ ($\overline{t}^{*} = 7.65$). A comparison of the two curves $\omega(t)$ at $k=0$ and at $k=4$ confirms that in the coupled problem ($k>0$), the diffusion coefficient $\overline{D} = (1 + k\omega)$ increases with the increase of the damage so that the concentration level increases at a greater speed and the time to fracture decreases.

7. Conclusion

Analysis of various approaches to simulation of the metal creep phenomenon shows that the most promising is the kinetic theory concept. This concept allows describing the characteristic features of the metal deformation under creep conditions up to the moment of fracture under various temperature-force loading programmes.

Certain above-mentioned advantages of the kinetic theory of creep and long-term strength compared to other theories should be emphasised. As described in this chapter, various versions of the kinetic theory can describe deformation and long-term fracture of metals under step loading. After the introduction of the structural kinetic parameter of damage, the kinetic theory makes it possible to take into account the non-monotonic dependence of the creep strain limit magnitude on the stress. Also, the kinetic theory allows considering the anisotropy of metal properties in simulating the long-term strength under complex stress state conditions. Moreover, this theory enables the researchers to take into account the influence of an aggressive medium on the creep and long-term strength of metals by introducing the kinetic parameter of aggressive medium concentration in the metal.

Apparently, it is not easy to highlight wide capabilities of the kinetic theory in one chapter. Therefore, this chapter naturally fails to show them all. However, the following promising research directions should be outlined:

1. Vibrocreep of metals under uniaxial and complex stress states
2. Dependence of the long-term strength under conditions of a biaxial stress state on the short-term loading programme
3. Simulation of a long-term fracture of a plate under a nonstationary complex stress state in the presence of an aggressive medium
4. Simulation of the blocking effect of the diffusion process.

Funding

The work was supported by the Russian Foundation for Fundamental Research (project No. 17-08-00210).

Author details

Alexander Lokoshchenko and Leonid Fomin*

*Address all correspondence to: fleonid1975@mail.ru

Research Institute of Mechanics, Lomonosov Moscow State University, Moscow, Russian Federation

References

[1] Kachanov LM. About time of rupture in the creep condition. Academy of Sciences of the USSR. Department of Tech. Sciences. 1958;(8):26-31. [in Russian]

[2] Rabotnov YN. On the mechanism of creep rupture. In: Questions of Strength of Materials and Structures. Moscow: Publishing House of the Academy of Sciences of the USSR; 1959, p. 5-7. [in Russian]

[3] Rabotnov YN. Creep of Structural Elements. Moscow: Nauka; 1966, 752 p. [in Russian]

[4] Rabotnov YN. Creep Problem in Structural Members. Amsterdam: North Holland; 1969

[5] Kachanov LM. The Theory of Creep. Moscow: Fizmatgiz; 1960, 456 p. [in Russian]

[6] Lokoshchenko AM. Creep and Creep Rupture of Metals. Moscow: Fizmatlit; 2016, 504 p. [in Russian]

[7] Lokoshchenko AM. Cambridge. In: Creep and Creep Rupture of Metals, International Science Publishing. 2017 (in print)

[8] Lokoshchenko AM. Use of a vector damage parameter in modeling of long-term strength of metals. Mechanics of Solids. 2016;**51**(3):315-320. DOI: 10.3103/S0025654416030080

[9] Sosnin OV, Gorev BV, Nikitenko AF. In: Lavrentieva MA, editor. Energy Version of the Theory of Creep. Novosibirsk: Institute of Hydrodynamics; 1986 96 p

[10] Vilesova NS, Namestnikov VS. About one hardening parameter. Journal of Applied Mechanics and Technical Physics. 1964;(3):43-47. [in Russian]

[11] Namestnikov VS, Rabotnov Yu N. On the hypothesis of the equation of state for creep. Journal of Applied Mechanics and Technical Physics. 1961;(3):101-102. [in Russian]

[12] Lokoshchenko AM, Shesterikov SA. Model of creep rupture with a non-monotonous dependence of deformation on stress. Journal of Applied Mechanics and Technical Physics. 1982;(1):160-163. [in Russian]

[13] Hayhurst DR. Creep rupture under multi-axial states of stress. Journal of the Mechanics and Physics of Solids. 1972;**20**(6):381-390

[14] Lokoshchenko AM, Myakotin EA, Shesterikov SA. Creep and long-term strength of steel Cr18Ni10Ti under conditions of a complex stress state. Academy of Sciences of the USSR. Mechanics of Solids. 1979;(4):87-94. [in Russian]

[15] Lokoshchenko AM. The application of an approximate analysis of the diffusion process for a description of creep and creep rupture. International Journal of Mechanical Science. 2005;**47**:359-373. DOI: 10.1016/j.ijmecsci.2005.01.011

[16] Fomin LV. Description of creep rupture strength of tensile rod with rectangular and circular cross-section at high temperature air media. Journal of Samara State Technical University, Series Physical and Mathematical Sciences. 2013;**3**(32):87-97

[17] Lokoshchenko AM, Fomin LV. Creep and long-term strength of rods under tension and bending in the presence of an aggressive medium. Izvestiya MSTU "MAMI". Natural Sciences. Scientific peer-reviewed journal. - M., Moscow State Technical University "MAMI", **2**(24), 2015, v. 4. 76 – 85

[18] Fomin LV. Steady-state creep of a composite rod in tension in the presence of an aggressive environment. Mechanics of Composite Materials. 2017;**52**(6):741-750. DOI: 10.1007/s11029-017-9624-5

Chapter 5

Kinetic Equations of Active Soft Matter

Viktor Gerasimenko

Additional information is available at the end of the chapter

http://dx.doi.org/10.5772/intechopen.70667

Abstract

We consider a new approach to the description of the collective behavior of complex systems of mathematical biology based on the evolution equations for observables of such systems. This representation of the kinetic evolution seems, in fact, the direct mathematically fully consistent formulation modeling the collective behavior of biological systems since the traditional notion of the state in kinetic theory is more subtle and it is an implicit characteristic of the populations of living creatures.

Keywords: kinetic equation, marginal observables, scaling limit, active soft matter

1. Introduction

The rigorous derivation of kinetic equations for soft condensed matter remains an open problem so far. It should be noted wide applications of these evolution equations to the description of collective processes of various nature [1–14], in particular, the collective behavior of complex systems of mathematical biology [13–23]. We emphasize that the considerable advance in solving the problem of rigorous modeling of the kinetic evolution of systems with a large number of constituents (entities) of mathematical biology, in particular, systems of large number of cells, is recently observed [20–26] (and see references cited therein).

In modern research, the main approach to the problem of the rigorous derivation of kinetic equation consists in the construction of scaling limits of a solution of evolution equations which describe the evolution of states of a many-particle system, in particular, a perturbative solution of the corresponding BBGKY hierarchy [2–4].

In this chapter, we review a new approach to the description of the collective behavior of complex systems of mathematical biology [17, 18] within the framework of the evolution of observables. This representation of the kinetic evolution seems, in fact, the direct mathematically fully consistent formulation modeling kinetic evolution of biological systems since the notion of the state is more subtle and it is an implicit characteristic of populations of living creatures.

One of the advantages of the developed approach is the opportunity to construct kinetic equations in scaling limits, involving initial correlations, in particular, that can characterize the condensed states of soft matter. We note also that such approach is also related to the problem of a rigorous derivation of the non-Markovian kinetic-type equations from underlying many-cell dynamics which make it possible to describe the memory effects of the kinetic evolution of cells.

Using suggested approach, we establish a mean field asymptotic behavior of the hierarchy of evolution equations for marginal observables of a large system of interacting stochastic processes of collisional kinetic theory [24], modeling the microscopic evolution of active soft condensed matter [14, 15]. The constructed scaling limit of a non-perturbative solution of this hierarchy is governed by the set of recurrence evolution equations, namely, by the dual Vlasov hierarchy for interacting stochastic processes.

Furthermore, we established that for initial states specified by means of a one-particle distribution function and correlation functions the evolution of additive-type marginal observables is equivalent to a solution of the Vlasov-type kinetic equation with initial correlations, and a mean field asymptotic behavior of non-additive-type marginal observables is equivalent to the sequence of explicitly defined correlation functions which describe the propagation of initial correlations of active soft condensed matter.

2. On collisional dynamics of active soft condensed matter and the evolution of marginal observables

The many-constituent systems of active soft condensed matter [14, 15] are dynamical systems displaying a collective behavior which differs from the statistical behavior of usual gases [2, 4]. In the first place, their own distinctive features are connected with the fact that their constituents (entities or self-propelled particles) show the ability to retain various complexity features [14–18]. To specify such nature of entities, we consider the dynamical system suggested in papers [13, 24, 29] which is based on the Markov jump processes that must represent the intrinsic properties of living creatures.

A description of many-constituent systems is formulated in terms of two sets of objects: observables and states. The functional of the mean value of observables defines a duality between observables and states and as a consequence there exist two approaches to the description of the evolution of such systems, namely in terms of the evolution equations for observables and for states. In this section, we adduce some preliminary facts about dynamics of finitely many entities of various subpopulations described within the framework of non-equilibrium grand canonical ensemble [2].

We consider a system of entities of various M subpopulations introduced in paper [24] in case of non-fixed, i.e., arbitrary, but finite average number of entities. Every i th entity is characterized by: $\boldsymbol{u}_i = (j_i, u_i) \in \mathcal{J} \times \mathcal{U}$, where $j_i \in \mathcal{J} \equiv (1, \dots, M)$ is a number of its subpopulation, and $u_i \in \mathcal{U} \subset \mathbb{R}^d$ is its microscopic state [24]. The stochastic dynamics of entities of various subpopulations is described by the semigroup $e^{t\Lambda} = \oplus_{n=0}^{\infty} e^{t\Lambda_n}$ of the Markov jump process defined on the space C_γ of sequences $b=(b_0, b_1, \dots, b_n, \dots)$ of measurable bounded functions $b_n(\mathbf{u}_1, \dots, \mathbf{u}_n)$ that are symmetric with respect to permutations of the arguments $\mathbf{u}_1, \dots, \mathbf{u}_n$ and equipped with the norm:

$$\|b\|_{C_\gamma} = \max_{n\geq 0} \frac{\gamma^n}{n!} \|b_n\|_{C_n} = \max_{n\geq 0} \frac{\gamma^n}{n!} \max_{j_1, \dots, j_n} \max_{u_1, \dots, u_n} |b_n(\mathbf{u}_1, \dots, \mathbf{u}_n)|,$$

where $\gamma<1$ is a parameter. The infinitesimal generator Λ_n of collisional dynamics (the Liouville operator of n entities) is defined on the subspace C_n of the space C_γ and it has the following structure [24]:

$$\begin{aligned}(\Lambda_n b_n)(\mathbf{u}_1, \dots, \mathbf{u}_n) &\doteq \sum_{m=1}^{M} \varepsilon^{m-1} \sum_{i_1\neq \dots \neq i_m=1}^{n} \Big(\Lambda^{[m]}(i_1, \dots, i_m) b_n\Big)(\mathbf{u}_1, \dots, \mathbf{u}_n) = \\ \sum_{m=1}^{M} \varepsilon^{m-1} \sum_{i_1\neq \dots \neq i_m=1}^{n} & a^{[m]}(\mathbf{u}_{i_1}, \dots, \mathbf{u}_{i_m}) \bigg(\int_{\mathcal{J}\times\mathcal{U}} A^{[m]}(\mathbf{v}; \mathbf{u}_{i_1}, \dots, \mathbf{u}_{i_m}) \times \\ & b_n(\mathbf{u}_1, \dots, \mathbf{u}_{i_1-1}, \mathbf{v}, \mathbf{u}_{i_1+1}, \dots \mathbf{u}_n) d\mathbf{v} - b_n(\mathbf{u}_1, \dots, \mathbf{u}_n) \bigg),\end{aligned} \tag{1}$$

where $\varepsilon>0$ is a scaling parameter [28], the functions $a^{[m]}(\mathbf{u}_{i_1}, \dots, \mathbf{u}_{i_m}), m\geq 1$, characterize the interaction between entities, in particular, in case of $m=1$ it is the interaction of entities with an external environment. These functions are measurable positive bounded functions on $(\mathcal{J}\times\mathcal{U})^n$ such that: $0\leq a^{[m]}(\mathbf{u}_{i_1}, \dots, \mathbf{u}_{i_m}) \leq a_*^{[m]}$, where $a_*^{[m]}$ is some constant. The functions $A^{[m]}(\mathbf{v}; \mathbf{u}_{i_1}, \dots, \mathbf{u}_{i_m}), m\geq 1$, are measurable positive integrable functions which describe the probability of the transition of the i_1 entity in the microscopic state u_{i_1} to the state v as a result of the interaction with entities in the states $u_{i_2}, \dots, u_{i_m}$ (in case of $m=1$ it is the interaction with an external environment). The functions $A^{[m]}(\mathbf{v}; \mathbf{u}_{i_1}, \dots, \mathbf{u}_{i_m}), m\geq 1$, satisfy the conditions: $\int_{\mathcal{J}\times\mathcal{U}} A^{[m]}(\mathbf{v}; \mathbf{u}_{i_1}, \dots, \mathbf{u}_{i_m}) d\mathbf{v} = 1, m\geq 1$. We refer to paper [24], where examples of the functions $a^{[m]}$ and $A^{[m]}$ are given in the context of biological systems.

In case of $M=1$ generator (1) has the form $\sum_{i_1=1}^{n} \Lambda_n^{[1]}(i_1)$ and it describes the free stochastic evolution of entities, i.e., the evolution of self-propelled particles. The case of $M=m\geq 2$ corresponds to a system with the m -body interaction of entities in the sense accepted in kinetic theory [30]. The m-body interaction of entities is the distinctive property of biological systems in comparison with many-particle systems, for example, gases of atoms with a pair-interaction potential.

On the space C_n the one-parameter mapping $e^{t\Lambda_n}$ is a bounded *-weak continuous semigroup of operators.

The observables of a system of a non-fixed number of entities of various subpopulations are the sequences $O=(O_0, O_1(\mathbf{u}_1), \dots, O_n(\mathbf{u}_1, \dots, \mathbf{u}_n), \dots)$ of functions $O_n(\mathbf{u}_1, \dots, \mathbf{u}_n)$ defined on $(\mathcal{J}\times\mathcal{U})^n$ and O_0 is a real number. The evolution of observables is described by the sequences $O(t)=(O_0, O_1(t, \mathbf{u}_1), \dots, O_n(t, \mathbf{u}_1, \dots, \mathbf{u}_n), \dots)$ of the functions

$$O_n(t) = e^{t\Lambda_n} O_n^0, \quad n\geq 1,$$

that is, they are the corresponding solution of the Cauchy problem of the Liouville equations (or the Kolmogorov forward equation) with corresponding initial data O_n^0:

$$\frac{\partial}{\partial t}O_n(t) = \Lambda_n O_n(t),$$
$$O_n(t)|_{t=0} = O_n^0, \quad n \geq 1,$$

or in case of n noninteracting entities (self-propelled particles) these equations have the form.

$$\frac{\partial}{\partial t}O_n(t,\mathbf{u}_1,\dots,\mathbf{u}_n) = \sum_{i=1}^{n} a^{[1]}(\mathbf{u}_i)\bigg(\int_{\mathcal{J}\times\mathcal{U}} A^{[1]}(\mathbf{v};\mathbf{u}_i)O_n(t,\mathbf{u}_1,\dots,\mathbf{u}_{i-1},\mathbf{v},\mathbf{u}_{i+1},\dots\mathbf{u}_n)d\mathbf{v}.$$
$$-O_n(t,\mathbf{u}_1,\dots,\mathbf{u}_n)\bigg), \quad n \geq 1.$$

The average values of observables (mean values of observables) are determined by the following positive continuous linear functional defined on the space C_γ:

$$\langle O\rangle(t) = (I,D(0))^{-1}(O(t),D(0)) \doteq (I,D(0))^{-1}\sum_{n=0}^{\infty}\frac{1}{n!}\int_{(\mathcal{J}\times\mathcal{U})^n} d\mathbf{u}_1\dots d\mathbf{u}_n\, O_n(t)\, D_n^0, \tag{2}$$

where $D(0) = (1, D_1^0, \dots, D_n^0, \dots)$ is a sequence of nonnegative functions D_n^0 defined on $(\mathcal{J}\times\mathcal{U})^n$ that describes the states of a system of a non-fixed number of entities of various subpopulations at initial time and $(I, D(0)) = \sum_{n=0}^{\infty}\frac{1}{n!}\int_{(\mathcal{J}\times\mathcal{U})^n} d\mathbf{u}_1\dots d\mathbf{u}_n\, D_n^0$ is a normalizing factor (the grand canonical partition function).

Let $L_\alpha^1 = \oplus_{n=0}^{\infty}\alpha^n L_n^1$ be the space of sequences $f=(f_0, f_1, \dots, f_n, \dots)$ of the integrable functions $f_n(\mathbf{u}_1, \dots, \mathbf{u}_n)$ defined on $(\boldsymbol{\mathcal{J}}\times\boldsymbol{\mathcal{U}})^n$, that are symmetric with respect to permutations of the arguments $\mathbf{u}_1, \dots, \mathbf{u}_n$, and equipped with the norm:

$$\|f\|_{L_\alpha^1} = \sum_{n=0}^{\infty}\alpha^n\|f_n\|_{L_n^1} = \sum_{n=0}^{\infty}\alpha^n\sum_{j_1\in\mathcal{J}}\dots\sum_{j_n\in\mathcal{J}}\int_{\mathcal{U}^n} du_1\dots du_n |f_n(\mathbf{u}_1,\dots,\mathbf{u}_n)|,$$

where $\alpha > 1$ is a parameter. Then for $D(0)\in L^1$ and $O(t)\in C_\gamma$ average value functional (2) exists and it determines a duality between observables and states.

As a consequence of the validity for functional (2) of the following equality:

$$(I,D(0))^{-1}(O(t),D(0)) = (I,D(0))^{-1}\left(e^{t\Lambda}O(0)\,D(0)\right) =$$
$$\left(I,e^{t\Lambda^*}D(0)\right)^{-1}\left(O(0)\,e^{t\Lambda^*}D(0)\right) \equiv (I,D(t))^{-1}(O(0),D(t)),$$

where $e^{t\Lambda^*} = \oplus_{n=0}^{\infty}e^{t\Lambda_n^*}$ is the adjoint semigroup of operators with respect to the semigroup $e^{t\Lambda} = \oplus_{n=0}^{\infty}e^{t\Lambda_n}$, it is possible to describe the evolution within the framework of the evolution of states. Indeed, the evolution of all possible states, i.e. the sequence $D(t)=(1, D_1(t,\mathbf{u}_1), \dots, D_n(t, \mathbf{u}_1, \dots, \mathbf{u}_n), \dots)\in L^1$ of the distribution functions $D_n(t), n\geq 1$, is determined by the formula:

$$D_n(t) = e^{t\Lambda_n^*} D_n^0, \quad n \geq 1,$$

where the generator Λ_n^* is the adjoint operator to operator (1) and on L_n^1 it is defined as follows:

$$(\Lambda_n^* f_n)(\mathbf{u}_1, \ldots, \mathbf{u}_n) \doteq \sum_{m=1}^{M} \varepsilon^{m-1} \sum_{i_1 \neq \ldots \neq i_m = 1}^{n} \Big(\int_{\mathcal{J}\times\mathcal{U}} A^{[m]}(\mathbf{u}_{i_1}; \mathbf{v}, \mathbf{u}_{i_2}, \ldots, \mathbf{u}_{i_m}) a^{[m]} \\ (\mathbf{v}, \mathbf{u}_{i_2}, \ldots, \mathbf{u}_{i_m}) f_n(\mathbf{u}_1, \ldots, \mathbf{u}_{i_1-1}, \mathbf{v}, \mathbf{u}_{i_1+1}, \ldots, \mathbf{u}_n) d\mathbf{v} - a^{[m]}(\mathbf{u}_{i_1}, \ldots, \mathbf{u}_{i_m}) f_n(\mathbf{u}_1, \ldots, \mathbf{u}_n) \Big), \tag{3}$$

where the functions $A^{[m]}$ and $a^{[m]}$ are defined as above in (1).

The function $D_n(t)$ is a solution of the Cauchy problem of the dual Liouville equation (or the Kolmogorov backward equation).

On the space L_n^1 the one-parameter mapping $e^{t\Lambda_n^*}$ is a bounded strong continuous semigroup of operators [26].

For the description of microscopic behavior of many-entity systems we also introduce the hierarchies of evolution equations for marginal observables and marginal distribution functions known as the dual BBGKY hierarchy and the BBGKY hierarchy, respectively [26]. These hierarchies are constructed as the evolution equations for one more method of the description of observables and states of finitely many entities.

An equivalent approach to the description of observables and states of many-entity systems is given in terms of marginal observables $B(t) = (B_0, B_1(t, \mathbf{u}_1), \ldots, B_s(t, \mathbf{u}_1, \ldots, \mathbf{u}_s), \ldots)$ and marginal distribution functions $F(0) = \big(1, F_1^{0,\varepsilon}(\mathbf{u}_1), \ldots, F_s^{0,\varepsilon}(\mathbf{u}_1, \ldots, \mathbf{u}_s), \ldots\big) \in L_\alpha^1$.

Considering formula (2), marginal observables and marginal distribution functions are introduced according to the equality:

$$\langle O \rangle(t) = (I, D(0))^{-1}(O(t), D(0)) = (B(t), F(0)),$$

where $(I, D(0))$ is a normalizing factor defined as above. If $F(0) \in L_\alpha^1$ and $B(0) \in C_\gamma$, then at $t \in \mathbb{R}$ the functional $(B(t), F(0))$ exists under the condition that: $\gamma > \alpha^{-1}$.

Thus, the relationship of marginal distribution functions $F(0) = \big(1, F_1^{0,\varepsilon}, \ldots, F_s^{0,\varepsilon}, \ldots\big)$ and the distribution functions $D(0) = \big(1, D_1^0, \ldots, D_n^0, \ldots\big)$ is determined by the formula:

$$F_s^{0,\varepsilon}(\mathbf{u}_1, \ldots, \mathbf{u}_s) \doteq (I, D(0))^{-1} \sum_{n=0}^{\infty} \frac{1}{n!} \int_{(\mathcal{J}\times\mathcal{U})^n} d\mathbf{u}_{s+1} \ldots d\mathbf{u}_{s+n} \, D_{s+n}^0(\mathbf{u}_1, \ldots, \mathbf{u}_{s+n}), \quad s \geq 1,$$

and, respectively, the marginal observables are determined in terms of observables as follows:

$$B_s(t, \mathbf{u}_1, \ldots, \mathbf{u}_s) \doteq \sum_{n=0}^{s} \frac{(-1)^n}{n!} \sum_{j_1 \neq \ldots \neq j_n = 1}^{s} O_{s-n}\big(t, (\mathbf{u}_1, \ldots, \mathbf{u}_s) \setminus (\mathbf{u}_{j_1}, \ldots, \mathbf{u}_{j_n})\big), \quad s \geq 1. \tag{4}$$

Two equivalent approaches to the description of the evolution of many interacting entities are the consequence of the validity of the following equality for the functional of mean values of marginal observables:

$$(B(t), F(0)) = (B(0), F(t)),$$

where $B(0) = \left(1, B_1^{0,\varepsilon}, \ldots, B_s^{0,\varepsilon}, \ldots\right)$ is a sequence of marginal observables at initial moment.

We remark that the evolution of many-entity systems is usually described within the framework of the evolution of states by the sequence $F(t) = (1, F_1(t, \mathbf{u}_1), \ldots, F_s(t, \mathbf{u}_1, \ldots, \mathbf{u}_s), \ldots)$ of marginal distribution functions $F_s(t, \mathbf{u}_1, \ldots, \mathbf{u}_s)$ governed by the BBGKY hierarchy for interacting entities [13, 24].

The evolution of a non-fixed number of interacting entities of various subpopulations within the framework of marginal observables (4) is described by the Cauchy problem of the dual BBGKY hierarchy [25]:

$$\frac{d}{dt} B(t) = \Lambda + \sum_{n=1}^{\infty} \frac{1}{n!} \underbrace{[\ldots[\Lambda, a^+], \ldots, a^+]}_{n-times} B(t), \tag{5}$$

$$B(t)|_{t=0} = B(0), \tag{6}$$

where on C_γ the operator a^+ (an analog of the creation operator) is defined as follows

$$\left(a^+ b\right)_s(\mathbf{u}_1, \ldots, \mathbf{us}) \doteq \sum_{j=1}^{s} b_{s-1}\left(\mathbf{u}_1, \ldots, \mathbf{u}_{j-1}, \mathbf{u}_{j+1}, \ldots, \mathbf{u}_s\right),$$

the operator $\Lambda = \oplus_{n=0}^{\infty} \Lambda_n$ is defined by (1), and the symbol $[\cdot, \cdot]$ denotes the commutator of operators.

In the componentwise form, the abstract hierarchy (5) has the form:

$$\begin{aligned} \frac{\partial}{\partial t} B_s(t, u_1, \ldots, u_s) = \Lambda_s B_s(t, u_1, \ldots, u_s) + \sum_{n=1}^{s} \frac{1}{n!} \sum_{k=n+1}^{s} \frac{1}{(k-n)!} \times \\ \times \sum_{j_1 \neq \ldots \neq j_k = 1}^{s} \varepsilon^{k-1} \Lambda^{[k]}(j_1, \ldots, j_k) \sum_{i_1 \neq \ldots \neq i_n \in (j_1, \ldots, j_k)} B_{s-n}(t, (u_1, \ldots, u_s) \backslash (u_{i_1}, \ldots, u_{i_n})), \end{aligned} \tag{7}$$

$$B_s(t, \mathbf{u}_1, \ldots, \mathbf{u}_s)|_{t=0} = B_s^{0,\varepsilon}(\mathbf{u}_1, \ldots, \mathbf{u}_s), \quad s \geq 1, \tag{8}$$

where the operators Λ_s and $\Lambda^{[k]}$ are defined by formulas (1) and the functions $B_s^{0,\varepsilon}, s \geq 1$, are scaled initial data.

A solution $B(t) = (B_0, B_1(t, \mathbf{u}_1), \ldots, B_s(t, \mathbf{u}_1, \ldots, \mathbf{u}_s), \ldots)$ of the Cauchy problem of recurrence evolution Eqs (7), (8) is given by the following expansions [26]:

$$B_s(t,\mathbf{u}_1,\dots,\mathbf{u}_s)=\sum_{n=0}^{s}\frac{1}{n!}\sum_{j_1\neq\dots\neq j_n=1}^{s}\mathfrak{A}_{1+n}(t,\{Y\backslash Z\},Z)\,B_{s-n}^{0,\varepsilon}$$
$$\big(u_1,\dots,\mathbf{u}_{j_1-1},\mathbf{u}_{j_1+1},\dots,\mathbf{u}_{j_n-1},\mathbf{u}_{j_n+1},\dots,\mathbf{u}_s\big),\quad s\geq 1, \tag{9}$$

where the $(1+n)th$-order cumulant of the semigroups $\{e^{t\Lambda_k}\}_{t\in\mathbb{R}}, k\geq 1$, is determined by the formula [25]:

$$\mathfrak{A}_{1+n}(t,\{Y\backslash Z\},Z)\doteq\sum_{\mathrm{P}:(\{Y\ Z\},Z)=\bigcup_i Z_i}(-1)^{|\mathrm{P}|-1}(|\mathrm{P}|-1)!\prod_{Z_i\subset\mathrm{P}}e^{t\Lambda_{|\theta(Z_i)|}}, \tag{10}$$

the sets of indexes are denoted by $Y\equiv(1,\dots,s)$, $Z\equiv(j_1,\dots,j_n)\subset Y$, the set $\{Y\setminus Z\}$ consists from one element $Y\setminus Z=(1,\dots,j_1-1,j_1+1,\dots,j_n-1,j_n+1,\dots,s)$ and the mapping $\theta(\cdot)$ is the declusterization operator defined as follows: $\theta(\{Y\setminus Z\},Z)=Y$.

The simplest examples of expansions for marginal observables (9) have the following form:

$$B_1(t,\mathbf{u}_1)=\mathfrak{A}_1(t,1)B_1^{\varepsilon,0}(\mathbf{u}_1),$$

$$B_2(t,\mathbf{u}_1,\mathbf{u}_2)=\mathfrak{A}_1(t,\{1,2\})B_2^{\varepsilon,0}(\mathbf{u}_1,\mathbf{u}_2)+\mathfrak{A}_2(t,1,2)\big(B_1^{\varepsilon,0}(\mathbf{u}_1)+B_1^{\varepsilon,0}(\mathbf{u}_2)\big),$$

and, respectively:

$$\mathfrak{A}_1(t,\{1,2\})=e^{t\Lambda_2(1,2)},$$

$$\mathfrak{A}_2(t,1,2)=e^{t\Lambda_2(1,2)}-e^{t\Lambda_1(1)}e^{t\Lambda_1(2)}.$$

For initial data $B(0)=\big(B_0,B_1^{0,\varepsilon},\dots,B_s^{0,\varepsilon},\dots\big)\in C_\gamma$ the sequence $B(t)$ of marginal observables given by expansions (9) is a classical solution of the Cauchy problem of the dual BBGKY hierarchy for interacting entities (7), (8).

We note that a one-component sequence of marginal observables corresponds to observables of certain structure, namely the marginal observable $B^{(1)}(0)=\big(0,b_1^{\varepsilon}(\mathbf{u}_1),0,\dots\big)$ corresponds to the additive-type observable, and a one-component sequence of marginal observables $B^{(k)}(0)=\big(0,\dots,0,b_k^{\varepsilon}(\mathbf{u}_1,\dots,\mathbf{u}_k),0,\dots\big)$ corresponds to the k-ary-type observable [25]. If in capacity of initial data (8) we consider the additive-type marginal observables, then the structure of solution expansion (9) is simplified and attains the form

$$B_s^{(1)}(t,\mathbf{u}_1,\dots,\mathbf{u}_s)=\mathfrak{A}_s(t,1,\dots,s)\sum_{j=1}^{s}b_1^{\varepsilon}(\mathbf{u}_j),\quad s\geq 1. \tag{11}$$

In the case of k-ary-type marginal observables solution expansion (9) has the form

$$B_s^{(k)}(t,u_1,\ldots,u_s)=\frac{1}{(s-k)!}\sum_{j_1\neq\ldots\neq j_{s-k}=1}^{s}\mathfrak{A}_{1+s-k}\big(t,\{(1,\ldots,s)\backslash(j_1,\ldots,j_{s-k})\},$$
$$(j_1,\ldots,j_{s-k})\big)\,b_k^{\varepsilon}\big(u_1,\ldots,u_{j_1-1},u_{j_1+1},\ldots,u_{j_s-k-1},u_{j_s-k+1},\ldots,u_s\big),\quad s\geq k, \tag{12}$$

and, if $1\leq s<k$, we have $B_s^{(k)}(t)=0$.

We remark also that expansion (9) can be also represented in the form of the perturbation (iteration) series [25] as a result of applying of analogs of the Duhamel equation to cumulants of semigroups of operators (10).

3. A mean field asymptotic behavior of the marginal observables and the kinetic evolution of states

To consider mesoscopic properties of a large system of interacting entities we develop an approach to the description of the kinetic evolution within the framework of the evolution equations for marginal observables. For this purpose we construct the mean field asymptotics [9] of a solution of the Cauchy problem of the dual BBGKY hierarchy for interacting entities, modeling of many-constituent systems of active soft condensed matter [26, 27].

We restrict ourself by the case of $M=2$ subpopulations to simplify the cumbersome formulas and consider the mean field scaling limit of non-perturbative solution (9) of the Cauchy problem of the dual BBGKY hierarchy for interacting entities (7), (8).

Let for initial data $B_s^{0,\varepsilon}\in C_s$ there exists the limit function $b_s^0\in C_s$

$$\mathrm{w}^*-\lim_{\varepsilon\to 0}\big(\varepsilon^{-s}B_s^{0,\varepsilon}-b_s^0\big)=0,\quad s\geq 1,$$

then for arbitrary finite time interval there exists a mean field limit of solution (9) of the Cauchy problem of the dual BBGKY hierarchy for interacting entities (7), (8) in the sense of the *-weak convergence of the space C_s

$$\mathrm{w}^*-\lim_{\varepsilon\to 0}\big(\varepsilon^{-s}B_s(t)-b_s(t)\big)=0,\quad s\geq 1,$$

where the limit sequence $b(t)=(b_0,b_1(t),\ldots,b_s(t),\ldots)$ of marginal observables is determined by the following expansions:

$$b_s(t,u_1,\ldots,u_s)=\sum_{n=0}^{s-1}\int_0^t dt_1\ldots\int_0^{t_{n-1}}dt_n\, e^{(t-t_1)\sum\limits_{k_1=1}^{s}\Lambda^{[1]}(k_1)}\sum_{i_1\neq j_1=1}^{s}\Lambda^{[2]}(i_1,j_1)e^{(t_1-t_2)\sum\limits_{l_1=1,l_1\neq j_1}^{s}\Lambda^{[1]}(l_1)}\ldots$$
$$e^{(t_{n-1}-t_n)\sum\limits_{k_n=1,\,k_n\neq(j_1,\ldots,j_{n-1})}^{s}\Lambda^{[1]}(k_n)}\sum_{\substack{i_n\neq j_n=1,\\ i_n,j_n\neq(j_1,\ldots,j_{n-1})}}^{s}\Lambda^{[2]}(i_n,j_n)e^{t_n\sum\limits_{l_n=1,\,l_n\neq(j_1,\ldots,j_n)}^{s}\Lambda^{[1]}(l_n)}$$
$$b_{s-n}^0\big((u_1,\ldots,u_s)\backslash(u_{j_1},\ldots,u_{j_n})\big),\quad s\geq 1. \tag{13}$$

In particular, the limit marginal observable $b_s^{(1)}(t)$ of the additive-type marginal observable (11) is determined as a special case of expansions (13):

$$b_s^{(1)}(t,\mathbf{u}_1,\dots,\mathbf{u}_s) = \int_0^t dt_1 \dots \int_0^{t_{s-2}} dt_{s-1}\, e^{(t-t_1)\sum_{k_1=1}^{s}\Lambda^{[1]}(k_1)} \sum_{i_1\neq j_1=1}^{s} \Lambda^{[2]}(i_1,j_1) e^{(t_1-t_2)\sum_{l_1=1,l_1\neq j_1}^{s}\Lambda^{[1]}(l_1)} \dots$$

$$e^{(t_{s-2}-t_{s-1})\sum_{k_{s-1}=1,\, k_n\neq (j_1,\dots,j_{n-1}))}^{s}\Lambda^{[1]}(k_n)} \sum_{\substack{i_{s-1}\neq j_{s-1}=1,\\ i_{s-1},j_{s-1}\neq (j_1,\dots,j_{s-2})}}^{s} \Lambda^{[2]}(i_{s-1},j_{s-1})\, e^{t_{s-1}\sum_{l_{s-1}=1,\, l_{s-1}\neq (j_1,\dots,j_{s-1}))}^{s}\Lambda^{[1]}(l_{s-1})} \times$$

$$b_1^0\big((\mathbf{u}_1,\dots,\mathbf{u}_s)\backslash(\mathbf{u}_{j_1},\dots,\mathbf{u}_{j_{s-1}})\big), \quad s\geq 1,$$

for example,

$$b_1^{(1)}(t,\mathbf{u}_1) = e^{t\Lambda^{[1]}(1)}\, b_1^0(\mathbf{u}_1),$$

$$b_2^{(1)}(t,\mathbf{u}_1,\mathbf{u}_2) = \int_0^t dt_1 \prod_{i=1}^{2} e^{(t-t_1)\Lambda^{[1]}(i)}\, \Lambda^{[2]}(1,2) \sum_{j=1}^{2} e^{t_1\Lambda^{[1]}(j)}\, b_1^0(\mathbf{u}_j).$$

The proof of this statement is based on the corresponding formulas for cumulants of asymptotically perturbed semigroups of operators (10).

If $b^0\in C_\gamma$, then the sequence $b(t)=(b_0, b_1(t), \dots, b_s(t), \dots)$ of limit marginal observables (13) is generalized global in time solution of the Cauchy problem of the dual Vlasov hierarchy:

$$\frac{\partial}{\partial t} b_s(t) = \sum_{j=1}^{s} \Lambda^{[1]}(j)\, b_s(t) + \sum_{j_1\neq j_2=1}^{s} \Lambda^{[2]}(j_1,j_2)\, b_{s-1}(t,\mathbf{u}_1,\dots,\mathbf{u}_{j_2-1},\mathbf{u}_{j_2+1},\dots,\mathbf{u}_s), \tag{14}$$

$$b_s(t,\mathrm{u}_1,\dots,\mathrm{u}_s)|_{t=0} = b_s^0(\mathrm{u}_1,\dots,\mathrm{u}_s), \quad s\geq 1, \tag{15}$$

where in recurrence evolution Eq. (14) the operators $\Lambda^{[1]}(j)$ and $\Lambda^{[2]}(j_1, j_2)$ are determined by Formula (1).

Further we consider initial states specified by a one-particle marginal distribution function in the presence of correlations, namely

$$f^{(c)} \equiv \left(1, f_1^0(\mathrm{u}_1), g_2(\mathrm{u}_1,\mathrm{u}_2)\prod_{i=1}^{2} f_1^0(\mathrm{u}_i), \dots, g_s(\mathrm{u}_1,\dots,\mathrm{u}_s)\prod_{i=1}^{s} f_1^0(\mathrm{u}_i), \dots\right), \tag{16}$$

where the bounded functions $g_s\equiv g_s(\mathbf{u}_1, \dots, \mathbf{u}_s), s\geq 2$, are specified initial correlations. Such assumption about initial states is intrinsic for the kinetic description of complex systems in condensed states.

If $b(t)\in C_\gamma$ and $f_1^0\in L^1(\mathcal{J}\times\mathcal{U})$, then under the condition that $\|f_1^0\|_{L^1(\mathcal{J}\times\mathcal{U})} < \gamma$, there exists a mean field scaling limit of the mean value functional of marginal observables and it is determined by the following series expansion:

$$\left(b(t),f^{(c)}\right)=\sum_{s=0}^{\infty}\frac{1}{s!}\int_{(\mathcal{J}\times\mathcal{U})^{s}}du_1\dots du_s\,b_s(t,u_1,\dots,u_s)g_s(u_1,\dots,u_s)\prod_{i=1}^{s}f_1^0(u_i).$$

Then for the mean-value functionals of the limit initial additive-type marginal observables, i.e. of the sequences $b^{(1)}(0)=(0,b_1^0(\mathbf{u}_1),0,\dots)$ [25], the following representation is true:

$$\left(b^{(1)}(t),f^{(c)}\right)=\sum_{s=0}^{\infty}\frac{1}{s!}\int_{(\mathcal{J}\times\mathcal{U})^{s}}d\mathbf{u}_1\dots d\mathbf{u}_s\,b_s^{(1)}(t,\mathbf{u}_1,\dots,\mathbf{u}_s)g_s(\mathbf{u}_1,\dots,\mathbf{u}_s)\prod_{i=1}^{s}f_1^0(\mathbf{u}_i)$$
$$=\int_{(\mathcal{J}\times\mathcal{U})}d\mathbf{u}_1\,b_1^0(\mathbf{u}_1)f_1(t,\mathbf{u}_1). \quad (17)$$

In equality (17) the function $b_s^{(1)}(t)$ is given by a special case of expansion (13), namely

$$b_s^{(1)}(t,\boldsymbol{u}_1,\dots,\boldsymbol{u}_s)=\int_0^t dt_1\dots\int_0^{t_{s-2}}dt_{s-1}\,e^{(t-t_1)\sum\limits_{k_1=1}^{s}\Lambda^{[1]}(k_1)}\sum_{i_1\neq j_1=1}^{s}\Lambda^{[2]}(i_1,j_1)\,e^{(t_1-t_2)\sum\limits_{l_1=1,\ l_1\neq j_1}^{s}\Lambda^{[1]}(l_1)}$$
$$\dots e^{(t_{s-2}-t_{s-1})\sum\limits_{k_{s-1}=1,\ k_{s-1}\neq(j_1,\dots,j_{s-2})}^{s}\Lambda^{[1]}(k_{s-1})}\sum_{\substack{i_{s-1}\neq\ j_{s-1}=1,\\ i_{s-1},j_{s-1}\neq\ (j_1,\dots,j_{s-2})}}^{s}\Lambda^{[2]}(i_{s-1},j_{s-1})$$
$$\times e^{t_{s-1}\sum\limits_{l_{s-1}=1,\ l_{s-1}\neq(j_1,\dots,j_{s-1})}^{s}\Lambda^{[1]}(l_{s-1})}\,b_1^0((\boldsymbol{u}_1,\dots,\boldsymbol{u}_s)\backslash(\boldsymbol{u}_{j_1},\dots,\boldsymbol{u}_{j_{s-1}})),\quad s\geq 1,$$

and the limit one-particle distribution function $f_1(t)$ is represented by the series expansion:

$$f_1(t,\mathbf{u}_1)=\sum_{n=0}^{\infty}\int_0^t dt_1\dots\int_0^{t_{n-1}}dt_n\int_{(\mathcal{J}\times\mathcal{U})^n}d\mathbf{u}_2\dots d\mathbf{u}_{n+1}\,e^{(t-t_1)\Lambda^{*[1]}(1)}\times$$
$$\times\Lambda^{*[2]}(1,2)\prod_{j_1=1}^{2}e^{(t_1-t_2)\Lambda^{*[1]}(j_1)}\dots\prod_{j_{n-1}=1}^{n}e^{(t_{n-1}-t_n)\Lambda^{*[1]}(j_{n-1})}\times \quad (18)$$
$$\times\sum_{i_n=1}^{n}\Lambda^{*[2]}(i_n,n+1)\prod_{j_n=1}^{n+1}e^{t_n\Lambda^{*[1]}(j_n)}g_{1+n}(\mathbf{u}_1,\dots,\mathbf{u}_{n+1})\prod_{i=1}^{n+1}f_1^0(\mathbf{u}_i),$$

where the operators $\Lambda^{*[i]}, i=1,2$, are adjoint operators (3) to the operators $\Lambda^{[i]}, i=1,2$ defined by formula (1), and on the space L_n^1 defined as follows:

$$\Lambda^{*[1]}(i)f_n(\mathbf{u}_1,\dots,\mathbf{u}_n)\doteq\int_{\mathcal{J}\times\mathcal{U}}A^{[1]}(\mathbf{u}_i;\mathbf{v})a^{[1]}(\mathbf{v})\times$$
$$\times f_n(\mathbf{u}_1,\dots,\mathbf{u}_{i-1},\mathbf{v},\mathbf{u}_{i+1},\dots,\mathbf{u}_n)d\mathbf{v}-a^{[1]}(\mathbf{u}_i)f_n(\mathbf{u}_1,\dots,\mathbf{u}_n),$$

$$\Lambda^{*[2]}(i,j)f_n(\mathbf{u}_1,\dots,\mathbf{u}_n)\doteq\int_{\mathcal{J}\times\mathcal{U}}A^{[2]}(\mathbf{u}_i;\mathbf{v},\mathbf{u}_j)a^{[2]}(\mathbf{v},\mathbf{u}_j)\times$$
$$\times f_n(\mathbf{u}_1,\dots,\mathbf{u}_{i-1},\mathbf{v},\mathbf{u}_{i+1},\dots,\mathbf{u}_n)d\mathbf{v}-a^{[2]}(\mathbf{u}_i,\mathbf{u}_j)f_n(\mathbf{u}_1,\dots,\mathbf{u}_n),$$

where the functions $A^{[m]},a^{[m]},m=1,2$, are defined above in formula (1).

For initial data $f_1^0 \in L^1(\mathcal{J} \times \mathcal{U})$ limit marginal distribution function (18) is the Vlasov-type kinetic equation with initial correlations:

$$\frac{\partial}{\partial t} f_1(t, \mathbf{u}_1) = \Lambda^{*[1]}(1) f_1(t, \mathbf{u}_1) + \int_{\mathcal{J} \times \mathcal{U}} d\mathbf{u}_2 \Lambda^{*[2]}(1,2) \prod_{i_1=1}^{2} e^{t\Lambda^{*[1]}(i_1)} g_2(\mathbf{u}_1, \mathbf{u}_2) \prod_{i_2=1}^{2} e^{-t\Lambda^{*[1]}(i_2)} f_1(t, \mathbf{u}_1) f_1(t, \mathbf{u}_2), \tag{19}$$

$$f_1(t, \mathbf{u}_1)\big|_{t=0} = f_1^0(\mathbf{u}_1), \tag{20}$$

where the function $g_2(\mathbf{u}_1, \mathbf{u}_2)$ is initial correlation function specified initial state (16).

For mean value functionals of the limit initial k-ary marginal observables, i.e. of the sequences $b^{(k)}(0) = \left(0, \ldots, 0, b_k^0(\mathbf{u}_1, \ldots, \mathbf{u}_k), 0, \ldots\right)$, the following representation is true:

$$\left(b^{(k)}(t), f^{(c)}\right) = \sum_{s=0}^{\infty} \frac{1}{s!} \int_{(\mathcal{J} \times \mathcal{U})^s} d\mathbf{u}_1 \ldots d\mathbf{u}_s \, b_s^{(k)}(t, \mathbf{u}_1, \ldots, \mathbf{u}_s) g_s(\mathbf{u}_1, \ldots, \mathbf{u}_s) \prod_{i=1}^{s} f_1^0(\mathbf{u}_i) =$$

$$= \frac{1}{k!} \int_{(\mathcal{J} \times \mathcal{U})^k} d\mathbf{u}_1 \ldots d\mathbf{u}_k \, b_k^0(\mathbf{u}_1, \ldots, \mathbf{u}_k) \times \prod_{i_1=1}^{k} e^{t\Lambda^{*[1]}(i_1)} g_k(\mathbf{u}_1, \ldots, \mathbf{u}_k) \prod_{i_2=1}^{k} e^{-t\Lambda^{*[1]}(i_2)} \prod_{i=1}^{k} f_1(t, \mathbf{u}_i), \ k \geq 2, \tag{21}$$

where the limit one-particle marginal distribution function $f_1(t, u_i)$ is determined by series expansion (18) and the functions $g_k(\mathbf{u}_1, \ldots, \mathbf{u}_k), k \geq 2$, are initial correlation functions specified initial state (16).

Hence in case of k-ary marginal observables the evolution governed by the dual Vlasov hierarchy (14) is equivalent to a property of the propagation of initial correlations (21) for the k-particle marginal distribution function or in other words mean field scaling dynamics does not create correlations.

In case of initial states of statistically independent entities specified by a one-particle marginal distribution function, namely $f^{(c)} \equiv \left(1, f_1^0(\mathbf{u}_1), \ldots, \prod_{i=1}^{s} f_1^0(\mathbf{u}_i), \ldots\right)$, the kinetic evolution of k-ary marginal observables governed by the dual Vlasov hierarchy means the property of the propagation of initial chaos for the k-particle marginal distribution function within the framework of the evolution of states [4], i.e. a sequence of the limit distribution functions has the form $f(t) \equiv \left(1, f_1(t, \mathbf{u}_1), \ldots, \prod_{i=1}^{s} f_1(t, \mathbf{u}_i), \ldots\right)$, where the one-particle distribution function $f_1(t)$ is governed by the Vlasov kinetic Eq. [26]

$$\frac{\partial}{\partial t} f_1(t, \mathbf{u}_1) = \Lambda^{*[1]}(1) f_1(t, \mathbf{u}_1) + \int_{\mathcal{J} \times \mathcal{U}} d\mathbf{u}_2 \Lambda^{*[2]}(1,2) f_1(t, \mathbf{u}_1) f_1(t, \mathbf{u}_2).$$

We note that, according to equality (21), in the mean field limit the marginal correlation functions defined as cluster expansions of marginal distribution functions [30, 33, 34] namely,

$$f_s(t, \mathbf{u}_1, \ldots, \mathbf{u}_s) = \sum_{\mathrm{P}:(u_1,\ldots,u_s)=\bigcup_i U_i} \prod_{U_i \subset \mathrm{P}} g_{|U_i|}(t, \mathbf{U}_i), \quad s \geq 1,$$

has the following explicit form [27]:

$$g_1(t, \mathbf{u}_1) = f_1(t, \mathbf{u}_1), \tag{22}$$

$$g_s(t, \mathbf{u}_1, \ldots, \mathbf{u}_s) = \prod_{i_1=1}^{s} e^{t\Lambda^{*[1]}(i_1)} \widetilde{g}_s(\mathbf{u}_1, \ldots, \mathbf{u}_s) \prod_{i_2=1}^{s} e^{-t\Lambda^{*[1]}(i_2)} \prod_{j=1}^{s} f_1(t, \mathbf{u}_j), \quad s \geq 2,$$

where for initial correlation functions (16) it is used the following notations:

$$\widetilde{g}_s(\mathbf{u}_1, \ldots, \mathbf{u}_s) = \sum_{\mathrm{P}:\,(\mathbf{u}_1, \ldots, \mathbf{u}_s) = \bigcup_i \mathbf{U}_i} \prod_{U_i \subset \mathrm{P}} g_{|U_i|}(\mathbf{U}_i),$$

the symbol $\sum_{\mathrm{P}}$ means the sum over possible partitions P of the set of arguments $(\mathbf{u}_1, \ldots, \mathbf{u}_s)$ on $|\mathrm{P}|$ non-empty subsets $\mathbf{U}_i$, and the one-particle marginal distribution function $f_1(t)$ is a solution of the Cauchy problem of the Vlasov-type kinetic equation with initial correlations (19), (20).

Thus, an equivalent approach to the description of the kinetic evolution of large number of interacting constituents in terms of the Vlasov-type kinetic equation with correlations (19) is given by the dual Vlasov hierarchy (14) for the additive-type marginal observables.

4. The non-Markovian generalized kinetic equation with initial correlations

Furthermore, the relationships between the evolution of observables of a large number of interacting constituents of active soft condensed matter and the kinetic evolution of its states described in terms of a one-particle marginal distribution function are discussed.

Since many-particle systems in condensed states are characterized by correlations we consider initial states specified by a one-particle marginal distribution function and correlation functions, namely

$$F^{(c)} = \left(1, F_1^{0,\varepsilon}(\mathbf{u}_1), g_2^{\varepsilon}(\mathbf{u}_1, \mathbf{u}_2) \prod_{i=1}^{2} F_1^{0,\varepsilon}(\mathbf{u}_i), \ldots, g_s^{\varepsilon}(\mathbf{u}_1, \ldots, \mathbf{u}_s) \prod_{i=1}^{s} F_1^{0,\varepsilon}(\mathbf{u}_i), \ldots\right). \tag{23}$$

If the initial state is completely specified by a one-particle distribution function and a sequence of correlation functions (23), then, using a non-perturbative solution of the dual BBGKY hierarchy (9), in [31, 32] it was proved that all possible states at the arbitrary moment of time can be described within the framework of a one-particle distribution function governed by the non-Markovian generalized kinetic equation with initial correlations, i.e. without any approximations like in scaling limits as above.

Indeed, for initial states (23) for mean value functional (4) the equality holds

$$\left(B(t), F^{(c)}\right) = (B(0), F(t|F_1(t))), \tag{24}$$

where $F(t\mid F_1(t))=(1,F_1(t),F_2(t\mid F_1(t)),\ldots,F_s(t\mid F_1(t)),\ldots)$ is a sequence of marginal functionals of the state with respect to a one-particle marginal distribution function

$$F_1(t,\mathbf{u}_1)=\sum_{n=0}^{\infty}\frac{1}{n!}\int\limits_{(\mathcal{J}\times\mathcal{U})^n} d\mathbf{u}_2\ldots d\mathbf{u}_{n+1}\mathfrak{A}^*_{1+n}(t,1,\ldots,n+1)g^{\varepsilon}_{n+1}(\mathbf{u}_1,\ldots,\mathbf{u}_{n+1})\prod_{i=1}^{n+1}F_1^{0,\varepsilon}(\mathbf{u}_i). \quad (25)$$

The generating operator $\mathfrak{A}^*_{1+n}(t)$ of series (25) is the $(1+n)$-order cumulant of the semigroups of operators $\{e^{t\Lambda^*_n}\}_{t\geq 0}, n\geq 1$.

The marginal functionals of the state is defined by the series expansions:

$$F_s(t,\mathbf{u}_1,\ldots,\mathbf{u}_s|F_1(t))\doteq\sum_{n=0}^{\infty}\frac{1}{n!}\int\limits_{(\mathcal{J}\times\mathcal{U})^n} d\mathbf{u}_{s+1}\ldots d\mathbf{u}_{s+n}\,\mathfrak{V}_{1+n}(t,\{Y\},X\backslash Y)\prod_{i=1}^{s+n}F_1(t,\mathbf{u}_i), \quad (26)$$

where the following notations used: $Y\equiv(1,\ldots,s)$, $X\setminus Y\equiv(s+1,\ldots,s+n)$ and the generating operators $\mathfrak{V}_{1+n}(t), n\geq 0$, are defined by the expansions [31]:

$$\begin{aligned}
&\mathfrak{V}_{1+n}(t,\{Y\},X\setminus Y)\doteq\sum_{k=0}^{n}(-1)^k\sum_{m_1=1}^{n}\ldots\sum_{m_k=1}^{n-m_1-\ldots-m_{k-1}}\frac{n!}{(n-m_1-\ldots-m_k)!}\\
&\times\widehat{\mathfrak{A}}_{1+n-m_1-\ldots-m_k}(t,\{Y\},s+1,\ldots,s+n-m_1-\ldots-m_k)\prod_{j=1}^{k}\sum_{k_2^j=0}^{m_j}\ldots\\
&\sum_{k^j_{n-m_1-\ldots-m_j+s}=0}^{k^j_{n-m_1-\ldots-m_j+s-1}}\prod_{i_j=1}^{s+n-m_1-\ldots-m_j}\frac{1}{\left(k^j_{n-m_1-\ldots-m_j+s+1-i_j}-k^j_{n-m_1-\ldots-m_j+s+2-i_j}\right)!}\\
&\times\widehat{\mathfrak{A}}_{1+k^j_{n-m_1-\ldots-m_j+s+1-i_j}-k^j_{n-m_1-\ldots-m_j+s+2-i_j}}(t,i_j,s+n-m_1-\ldots-m_j+1\\
&+k^j_{s+n-m_1-\ldots-m_j+2-i_j},\ldots,s+n-m_1-\ldots-m_j+k^j_{s+n-m_1-\ldots-m_j+1-i_j}),
\end{aligned} \quad (27)$$

where $k_1^j\equiv m_j$, $k^j_{n-m_1-\ldots-m_j+s+1}\equiv 0$ and the evolution operators $\widehat{\mathfrak{A}}_n(t), n\geq 1,$ are cumulants of the semigroups of scattering operators $\left\{e^{t\Lambda^*_k}g^{\varepsilon}_k\prod_{i=1}^{k}e^{-t\Lambda^{*[1]}(i)}\right\}_{t\geq 0}, k\geq 1$. We adduce some examples of evolution operators (27):

$$\mathfrak{V}_1(t,\{Y\})=\widehat{\mathfrak{A}}_1(t,\{Y\})\doteq e^{t\Lambda^*_s}g^{\varepsilon}_s\prod_{i=1}^{s}e^{-t\Lambda^{*[1]}(i)},$$

$$\mathfrak{V}_2(t,\{Y\},s+1)=\widehat{\mathfrak{A}}_2(t,\{Y\},s+1)-\widehat{\mathfrak{A}}_1(t,\{Y\})\sum_{i_1=1}^{s}\widehat{\mathfrak{A}}_2(t,i_1,s+1).$$

If $\|F_1(t)\|_{L^1(\mathcal{J}\times\mathcal{U})}<e^{-(3s+2)}$, then for arbitrary $t\in\mathbb{R}$ series expansion (26) converges in the norm of the space L^1_s [30].

The proof of equality (24) is based on the application of cluster expansions to generating operators (10) of expansions (9) which are dual to the kinetic cluster expansions introduced in paper [35]. Then the adjoint series expansion can be expressed in terms of one-particle distribution function (25) in the form of the functional from the right-hand side of equality (24).

We emphasize that marginal functionals of the state (26) characterize the processes of the creation of correlations generated by dynamics of many-constituent systems of active soft condensed matter and the propagation of initial correlations.

For small initial data $F_1^{0,\varepsilon} \in L^1(\mathcal{J} \times \mathcal{U})$ [31], series expansion (25) is a global in time solution of the Cauchy problem of the generalized kinetic equation with initial correlations:

$$\frac{\partial}{\partial t} F_1(t, \mathbf{u}_1) = \Lambda^{*[1]}(1) F_1(t, \mathbf{u}_1) + \sum_{k=1}^{M-1} \frac{\varepsilon^k}{k!} \int\limits_{(\mathcal{J} \times \mathcal{U})^k} d\mathbf{u}_2 \ldots d\mathbf{u}_{k+1} \sum_{\substack{j_1 \neq \ldots \neq j_{k+1} \in \\ \in (1, \ldots, k+1)}} \Lambda^{*[k+1]}(j_1, \ldots, j_{k+1}) F_{k+1}(t, \mathbf{u}_1, \ldots, \mathbf{u}_{k+1} | F_1(t)), \tag{28}$$

$$F_1(t, \mathbf{u}_1)\big|_{t=0} = F_1^{0,\varepsilon}(\mathbf{u}_1). \tag{29}$$

For initial data $F_1^{0,\varepsilon} \in L^1(\mathcal{J} \times \mathcal{U})$ it is a strong (classical) solution and for an arbitrary initial data it is a weak (generalized) solution.

In particular case $M = 2$ of two subpopulations kinetic Eq. (28) has the following explicit form:

$$\frac{\partial}{\partial t} F_1(t, u_1) = \int\limits_{\mathcal{J} \times \mathcal{U}} A^{[1]}(u_1; v) a^{[1]}(v) F_1(t, v) dv - a^{[1]}(u_1) F_1(t, u_1) +$$

$$\int\limits_{\mathcal{J} \times \mathcal{U}} du_2 \left(\int\limits_{\mathcal{J} \times \mathcal{U}} A^{[2]}(u_1; v, u_2) a^{[2]}(v, u_2) F_2(t, v, u_2 | F_1(t)) dv - a^{[2]}(u_1, u_2) F_2(t, u_1, u_2 | F_1(t)) \right),$$

where the functions $A^{[k]}$ and $a^{[k]}$ are defined above.

We note that for initial states (23) specified by a one-particle (marginal) distribution function, the evolution of states described within the framework of a one-particle (marginal) distribution function governed by the generalized kinetic equation with initial correlations (28) is dual to the dual BBGKY hierarchy for additive-type marginal observables with respect to bilinear form (2), and it is completely equivalent to the description of states in terms of marginal distribution functions governed by the BBGKY hierarchy of interacting entities.

Thus, the evolution of many-constituent systems of active soft condensed matter described in terms of marginal observables in case of initial states (23) can be also described within the framework of a one-particle (marginal) distribution function governed by the non-Markovian generalized kinetic equation with initial correlations (28).

We remark, considering that a mean field limit of initial state (23) is described by sequence (16), a mean field asymptotics of a solution of the non-Markovian generalized kinetic equation with initial correlations (28) is governed by the Vlasov-type kinetic equation with initial correlations (19) derived above from the dual Vlasov hierarchy (14) for limit marginal observables of interacting entities [27]. Moreover, a mean field asymptotic behavior of marginal functionals of the state (26) describes the propagation in time of initial correlations like established property (22).

5. Conclusion

We considered an approach to the description of kinetic evolution of large number of interacting constituents (entities) of active soft condensed matter within the framework of the evolution of marginal observables of these systems. Such representation of the kinetic evolution seems, in fact, the direct mathematically fully consistent formulation modeling the collective behavior of biological systems since the notion of state is more subtle and implicit characteristic of living creatures.

A mean field scaling asymptotics of non-perturbative solution (9) of the dual BBGKY hierarchy (7) for marginal observables was constructed. The constructed scaling limit of a non-perturbative solution (9) is governed by the set of recurrence evolution equations (14), namely, by the dual Vlasov hierarchy for interacting stochastic processes modeling large particle systems of active soft condensed matter.

We established that the limit additive-type marginal observables governed by the dual Vlasov hierarchy (14) gives an equivalent approach to the description of the kinetic evolution of many entities in terms of a one-particle distribution function governed by the Vlasov kinetic equation with initial correlations (19). Moreover, the kinetic evolution of non-additive-type marginal observables governed by the dual Vlasov hierarchy means the property of the propagation of initial correlations (22) within the framework of the evolution of states.

One of the advantages of suggested approach in comparison with the conventional approach of the kinetic theory [2, 3, 4] is the possibility to construct kinetic equations in various scaling limits in the presence of initial correlations which can characterize the analogs of condensed states of many-particle systems of statistical mechanics for interacting entities of complex biological systems.

We note that the developed approach is also related to the problem of a rigorous derivation of the non-Markovian kinetic-type equations from underlying many-entity dynamics which make it possible to describe the memory effects of collective dynamics of complex systems modeling active soft condensed matter.

In case of initial states completely specified by a one-particle distribution function and correlations (23), using a non-perturbative solution of the dual BBGKY hierarchy (9), it was proved that all possible states at the arbitrary moment of time can be described within the framework of a one-particle distribution function governed by the non-Markovian generalized kinetic equation with initial correlations (28), i.e. without any approximations. A mean field asymptotics of a solution of kinetic equation with initial correlations (28) is governed by the Vlasov-type kinetic equation with initial correlations (19) derived above from the dual Vlasov hierarchy (14) for limit marginal observables.

Moreover, in the case under consideration the processes of the creation of correlations generated by dynamics of large particle systems of active soft condensed matter and the propagation of initial correlations are described by the constructed marginal functionals of the state (26) governed by the non-Markovian generalized kinetic equation with initial correlations (28).

Author details

Viktor Gerasimenko

Address all correspondence to: gerasym@imath.kiev.ua

Institute of Mathematics of the NAS of Ukraine, Kyiv, Ukraine

References

[1] Gallagher I, Saint-Raymond L, Texier B. From Newton to Boltzmann: Hard Spheres and Short-Range Potentials. EMS Publishing House: Zürich Lectures in Advanced Mathematics: Zürich: EMS, 2014

[2] Cercignani C, Gerasimenko VI, Petrina DY. Many-Particle Dynamics and Kinetic Equations. Springer: the Netherlands, 2012

[3] Villani C. A review of mathematical topics in collisional kinetic theory. In: Handbook of Mathematical Fluid Dynamics. Vol. 1. North-Holland, Amsterdam; 2002. p. 71

[4] Cercignani C, Illner R, Pulvirenti M. The Mathematical Theory of Dilute Gases. Berlin: Springer-Verlag; 1994

[5] Ukai S, Yang T. Mathematical Theory of Boltzmann Equation. Lecture Notes, Series-No. 8. Liu Bie Ju Centre for Math. Sci., City University of Hong Kong; Hong Kong, 2006

[6] Grad H. Principles of the kinetic theory of gases. In: Handbuch der Physik. Vol. 12. Berlin: Springer; 1958. p. 205

[7] Gerasimenko VI, Kornienko AG. The Boltzmann kinetic equation with correlations for hard sphere fluids. Reports of NAS of Ukraine. 2015;**17**(3)

[8] Saint-Raymond L. Kinetic models for superfluids: A review of mathematical results. Comptes Rendus Physique. 2004;**5**:65

[9] Golse F. On the dynamics of large particle systems in the mean field limit. In: Macroscopic and Large Scale Phenomena: Coarse Graining, Mean Field Limits and Ergodicity, Lect. Notes Appl. Math. Mech. Vol. 3. Springer; 2016. p. 1-144

[10] Pezzotti F, Pulvirenti M. Mean-field limit and semiclassical expansion of quantum particle system. Annales Henri Poincaré. 2009;**10**:145

[11] Bao W, Cai Y. Mathematical theory and numerical methods for Bose–Einstein condensation. Kinetic and Related Models. 2013;**6**(1):1

[12] Erdös L, Schlein B, Yau H-T. Derivation of the cubic nonlinear Schrödinger equation from quantum dynamics of many-body systems. Inventiones Mathematicae. 2007;**167**(3):515

[13] Lachowicz M. Links between microscopic and macroscopic descriptions. In: Multiscale Problems in the Life Sciences. From Microscopic to Macroscopic, Lecture Notes in Math. Vol. 1940; 2008. p. 201

[14] Bellouquid A, Delitala M. Mathematical Modeling of Complex Biological Systems: A Kinetic Theory Approach. Birkhäuser: Boston, 2006

[15] Marchetti MC, Joanny JF, Ramaswamy S, Liverpool TB, Prost J, Rao M, Aditi Simha R. Hydrodynamics of soft active matter. Reviews of Modern Physics. 2013;**85**:1143

[16] Bianca C. Thermostatted kinetic equations as models for complex systems in physics and life sciences. Physics of Life Reviews. 2012;**9**(4):359

[17] Menzel AM. Tuned, driven, and active soft matter. Physics Reports. 2015;**554**:1

[18] Vicsek T, Zafeiris A. Collective motion. Physics Reports. 2012;**517**:71

[19] Lachowicz M, Miękisz J. From Genetics to Mathematics. New Jersey: World Science; 2009

[20] Mones E, Czirók A, Vicsek T. Anomalous segregation dynamics of self-propelled particles. New Journal of Physics. 2015;**17**:063013

[21] Bellomo N, Carbonaro B. Toward a mathematical theory of living systems focusing on developmental biology and evolution: A review and perspectives. Physics of Life Reviews. 2011;**8**(1):1

[22] Bellomo N, Dogbé C. On the modeling of trafic and crowds: A survey of models, speculations and perspectives. SIAM Review. 2011;**53**:409

[23] Carlen E, Degond P, Wennberg B. Kinetic limits for pair-interaction driven master equations and biological swarm models. Mathematical Models and Methods in Applied Sciences. 2013;**23**:1339

[24] Lachowicz M. Individually-based Markov processes modeling nonlinear systems in mathematical biology. Nonlinear Analysis: Real World Applications. 2011;**12**:2396

[25] Borgioli G, Gerasimenko V. Initial-value problem of the quantum dual BBGKY hierarchy. Nuovo Cimento della Societa Italiana di Fisica C. 2010;**33**:71

[26] Gerasimenko VI, Fedchun Y. On kinetic models for the evolution of many-entity systems in mathematical biology. Journal of Coupled Systems and Multiscale Dynamics. 2013;**1**(2):273

[27] Gerasimenko VI, Fedchun Y. On semigroups of large particle systems and their scaling asymptotic behavior. In: Semigroups of Operators – Theory and Applications. Series: Springer Proceedings in Mathematics and Statistics. Vol. 113. Springer; Switzerland 2015. p. 165

[28] Banasiak J, Lachowicz M. Methods of Small Parameter in Mathematical Biology. Boston: Birkhäuser; 2014

[29] Lachowicz M, Pulvirenti M. A stochastic particle system modeling the Euler equation. Archive for Rational Mechanics and Analysis. 1990;**109**:81

[30] Gerasimenko VI. Hierarchies of quantum evolution equations and dynamics of many-particle correlations. In: Statistical Mechanics and Random Walks: Principles, Processes and Applications. N.Y: Nova Science Publ., Inc.; 2012. p. 233

[31] Gerasimenko VI, Fedchun YY. Kinetic equations of soft active matter. Reports of NAS of Ukraine. 2014;**11**(5)

[32] Gerasimenko VI, Tsvir ZA. On quantum kinetic equations of many-particle systems in condensed states. Physica A: Statistical Mechanics and its Applications. 2012;**391**(24):6362

[33] Gerasimenko VI, Polishchuk DO. Dynamics of correlations of Bose and Fermi particles. Mathematical Methods in the Applied Sciences. 2011;**34**(1):76

[34] Gerasimenko VI. The evolution of correlation operators of large particle quantum systems. Methods of Functional Analysis and Topology. 2017;**23**(2):123

[35] Gerasimenko VI, Gapyak IV. Hard sphere dynamics and the Enskog equation. Kinetic and Related Models. 2012;**5**(3):459

Chapter 6

Non-Linear Kinetic Analysis of Protein Assembly Based on Center Manifold Theory

Tatsuaki Tsuruyama

Additional information is available at the end of the chapter

http://dx.doi.org/10.5772/intechopen.70750

Abstract

This review introduces a novel mathematical description of protein assembly. Protein assembly occurs in a substantially open non-equilibrium and non-linear kinetic system. The goal of systems biology is to make predictions about such complicated systems, but few have conducted stability analysis for given systems. Particularly, simulated dynamic behaviors have not been sufficiently verified by kinetic analysis in predicting macromolecular protein interactions and assembly. The non-linearity of protein assembly kinetics is complex, and it is very difficult to determine a model of multi-protein interactions based on numerical calculation. We studied the non-linear kinetics involved in the diffusion process of proteins consisting of two or three species of macromolecules and set a novel model in which non-linearity is given by the diffusion coefficient that depends on the protein concentration. By making the diffusion coefficient concentration-dependent, non-linearity leads to a simple system model. Protein assembly is initiated by monomeric protein interactions and regulated by cofactors such as guanidine triphosphate (GTP) or adenosine triphosphate (ATP) binding to the monomer. This cofactor concentration promotes the dynamic behavior of protein assembly and can be treated as an order parameter. Further, kinetic stability analysis in the center manifold theory (CMT) is introduced for analyzing the behavior of the system around the critical state. Although CMT has not been sufficiently applied for stability analysis of protein assembly systems, this theory predicts the dynamic behavior of the assembly system around the critical point using concentration as a cofactor. Protein assembly theory will provide a novel framework for nonlinear multi-parametric analysis.

Keywords: protein assembly, center manifold theory, tubulin, non-linear kinetics, non-equilibrium state, diffusion coefficient, oscillation

1. Background

Protein assembly is essential for cellular activities such as cell signaling, gene expression by transcription factor complexes, cytoskeleton formation, endocytosis, and cell motility. This reaction system is one of the complicated systems and non-linear kinetics has been applied for understanding the dynamic behavior [1–8]. Among the protein assembly, tubulin and actin polymerization are well-known events that have been analyzed using numerous methods [9–13]. Microtubule and actin filaments consist of monomers that bind the cofactors guanidine triphosphate (GTP) or adenosine triphosphate (ATP) to acquire assembly activity [14–16]. In general, a protein interaction is controlled by an electric charge on the amino acid residue(s) of the protein, such as tyrosine, serine, and threonine, by covalently binding to or reacting with a cofactor such as ATP or GTP; subsequently, the monomer loses its interaction activity by hydrolyzing ATP or GTP into ADP or GDP or through dephosphorylation, which is mediated by the other protein's phosphatase activity or its own enzymatic activity [17, 18].

In particular, *dynamic instability* in tubulin polymerization has been extensively investigated. *Dynamic instability* signifies the intermittent transition between slow growth and rapid shrinkage in polymeric assemblies of microtubules [9–13, 19]. Further, intra-polymeric Brownian motion and fluctuation influence the structure and elasticity of tubules [20]. Zapperi and Mahadevan presented an excellent model in which the ratio of longitudinal to lateral interactions characterizes the assembly [21]. Hammele et al. presented a physical model and suggested the physical properties of the microtubules [22]. Nucleation is the rate-limiting step controlling the overall polymerization process [18]. The stable nucleus for polymerization is oligomers, and the growth of aggregates through elongation/dissociation follows. For stable growth, tubule lifespan is controlled by a GTP-cap that forms at their ends [19].

As another example of protein interactions, in the mitogen-activated protein kinase (MAPK) signaling cascade, a set of protein kinases and protein substrates construct the signaling network. The cofactor ATP/GTP transfers biological information in the reaction network to alter gene expression [6–8]. Mathematical models of this cascade have demonstrated that the system can act as an ultra-sensitive switch based on a combination of phosphorylation of protein substrates and implicit feedback, leading to multi-stability [23, 24, 25]. Recently, Ueno et al. reported that a model of MAPK signaling cascade functions as a band-path filtering system.

2. Protein interaction kinetics

2.1. Protein interaction model

Steps in protein assembly of microtubular polymerization are summarized as follows: (i) the protein achieves an interaction active state by reversibly binding to a cofactor, which provides the protein with assembly or interaction activity; the protein interaction activity decreases when a hydrolyzed inactive cofactor is bound compared to an active cofactor; (ii) the protein can hydrolyze the cofactor; (iii) the protein can exchange the inactive cofactor, such as ADP, with an active cofactor; and (iv) active cofactors are supplied continuously and externally. Thus, the interaction activity is self-limiting, in which the protein itself limits assembly activity, resulting in dynamic instability [10–13, 19, 23, 26–28].

Let us consider a three-monomer model in which an active cofactor-binding protein (X), oligomeric protein (W), and inactive cofactor-binding protein (Z) coexist. The sum of monomeric proteins is kept constant; this kinetics is attributed to a two-parametric analysis. First, X can reversibly associate with the oligomer:

$$X + W \rightarrow W(\text{kinetic rate coefficients} \rightarrow k_1) \tag{1}$$

Subsequently, Z is released from the oligomer:

$$W \rightarrow Z \tag{2}$$

Here, Pi signifies the released inorganic phosphate. Finally, in the presence of sufficient active cofactor, Z releases the inactive cofactor, ADP or GDP (I), binds an active cofactor, ATP or GTP (P), and recovers its interaction activity, returning to the protein

$$Z + P \rightarrow X + I \quad (k_3) \tag{3}$$

In addition, a monomeric protein has the potential to hydrolyze the cofactor by interaction:

$$X + X \rightarrow X + Z \quad (k_4) \tag{4}$$

$$X + Z \rightarrow 2Z \quad (k_5) \tag{5}$$

Formula (5) represents a self-reproducing reaction, which yields non-linear kinetics in this reaction system.

2.2. Concentration dependence of protein diffusion

Diffusion of proteins plays an important role in protein interactions. Analysis of dilute solutions of a macromolecule requires a greater understanding of the concentration dependence of the diffusion rate because of the hydrodynamics of protein solutions involving mutual diffusion of protein molecules. One of the approaches is applicable to the study of self-diffusion in solutions [29–32]. In fact, proteins interact or associate with other monomeric proteins and phosphorylate or are phosphorylated by the proteins. In dilute solution, proteins may diffuse in a free manner with sufficiently large vacant space that accounts for only a fraction of the volume of a protein molecule. These vacancies are sufficiently large to be occupied by proteins that are as large as the hydrodynamic volume. The effects of molecular shape and size, solvent, and environment such as ion intensity and pH determine the concentration dependence of the diffusion rate. Here, the diffusion rate D_0 is set as the probability of vacancy formation and does not depend on the velocity at which a monomer diffuses. The probability P is given by a void adjacent to the objective protein, which is sufficiently large to permit diffusion: $D = D_oP$.

The probability $P(V)$ of forming a vacancy of volume V in the solution is given by

$$P(V) = D_0 \exp\left[-\beta c \rho V_e/(1 - cV_e)\right] \tag{6}$$

where V_e is protein exclusion volume, c is the protein solution concentration, and β is a constant that reflects the effects of interactions with other macromolecules and shape on the probability of vacancy formation. Thus, D depends on the size and shape of the protein and likely other factors

$$D = D_0 \int_{V_e}^{\infty} p(V)dV = D_0 \exp\left[-\beta c \rho V_e/(1 - cV_e)\right] \tag{7}$$

At low concentrations of the macromolecule

$$D = D_0\left(1 - \beta c \rho V_e/(1 - cV_e)\right) \simeq D_0\left(1 + (1 - \beta)cV_e\right) \tag{8}$$

Thus, the dependence of the diffusion coefficient can be described using the protein solution concentration (**Figure 1**).

2.3. Viscosity and diffusion coefficient of protein

The compatibility of Eq. (8) with the experimental data strongly suggests that the concentration dependence of the protein diffusion constant is governed by excluded volume interactions, which may be predicted by calculating equilibrium protein density fluctuations. Eq. (8) is consistent with the equation describing the viscosity η of concentrated protein solutions

$$\eta/\eta_0 = \exp\left(\rho(\eta)/\left(1 - \rho(\eta)(k/v)\right)\right) \tag{9}$$

where η_0 is the solution viscosity at infinite dilution, η is the intrinsic viscosity of the solution, c is the protein solution density, and k/v is a constant that corrects for the overlap of free volume v.

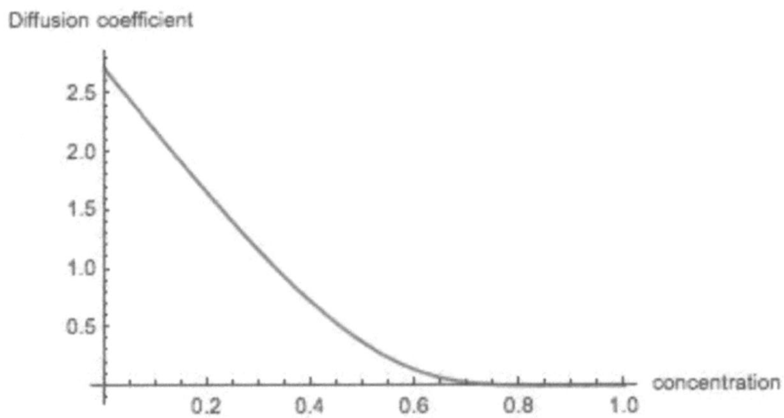

Figure 1. Self-diffusion rate constants D for concentration of protein, c, with theoretical curve from Eq. (8) with $\beta = 3.0$. Near the point $x \sim 0$, the diffusion coefficient obeys $D = 1 - 2c$. The vertical axis represents the diffusion coefficient $D \times 10^7$ (cm^2/s) and horizontal axis represents hemoglobin concentration (g/dL). The graph is shown using new arbitrary values with reference to experimental data.

2.4. Diffusion-limited protein association

Because protein diffusion in the viscous cytoplasm is significantly slower than other reactions between interactions with other molecules, the protein assembly reaction is a diffusion-controlled or diffusion-limit process. In general, the diffusion rate of monomer is given by the concentration gradient that is the molar flux per unit area, J. This is the additional rate of monomer X towards oligomer W multiplied by the area of the spherical surface of radius R of the reactive end of polymer or oligomer

$$\text{Rate of reaction} = 4\pi R^2 J \tag{10}$$

In Fick's first law, the flux towards X is proportional to the concentration gradient at radius of macromolecule R

$$J_X = D_X \left(\frac{d[X]_r}{dr} \right)_{at\ \mathrm{r=R}} = D_X \frac{[X]}{R} \tag{11}$$

By substitution of Eq. (10) into Eq. (11), the rate of reaction v is given as

$$v = 4\pi R D_X X^X \tag{12}$$

The rate of the diffusion-controlled reaction is equal to the average flow of X molecule to all W molecules. Accordingly, the global flow of all X to W is $4\pi\ R^* D_X N_A\ XW$. Similarly, the flow of all W to X is $4\pi\ R D_W N_A\ WX$. Further, using the sum of the diffusion coefficients of the two species, the diffusion coefficient is rewritten as $D = (D_X + D_W)/2$. Then, the addition rate of X to W is given by

$$\text{Addition rate} = 4\pi R\ k_X{}^o DXW \tag{13}$$

In reality, the diffusion rate of oligomer is negligible relative to that of monomer, and the addition rate is given as

$$\text{Addition rate} = 4\pi R\ k_X{}^o D_X XW = k_X D_X XW \tag{14}$$

and

$$k_X = 4\pi R\ k_X{}^o \tag{15}$$

Using the above formula, the kinetic rate of c_i and items of interaction between monomers is as follows [8]:

$$\frac{dc_i}{dt} = k_j D_i c_i + f(c_i) \tag{16}$$

where $f(c_j)$ represents the sum of the reaction kinetic items of jth species, c_j, except the diffusion process. When the potential of the electric field, $\phi(r)$, is considered, the flux is described using spatial coordinate r

$$J = -D_i\left[\frac{dc_j}{dr} + \frac{1}{k_BT}c_j\frac{d\phi(r)}{dr}\right] \tag{17}$$

In the solvent including two chemical species, the relative movement is related to the local concentration gradient of c_j and potential energy. When the gradient is described using Eq. (17), monomer X moves across the sphere with radius R surrounding W and can be described by

$$J = -(D_X + D_W)\left[\frac{dX}{dr} + \frac{1}{k_BT}X\frac{d\phi(r)}{dr}\right] \tag{18}$$

when X and the active site on the oligomer or polymer W interact to assemble or elongate the polymer, which is determined by R. The total flux will be equivalent to the chemical reaction rate using an arbitrary kinetic coefficient k

$$\frac{dX}{dt} = -kXW = -4\pi R^2 J \tag{19}$$

At the steady state, flux across the sphere with a radius, or a shape parameter, r is constant for any values. Accordingly, R in Eq. (19) is replaced with r using Eq. (18)

$$\frac{dX}{dt} = -4\pi R^2(D_X + D_W)\left[\frac{dX}{dr} + \frac{1}{k_BT}X\frac{d\phi(r)}{dr}\right] \tag{20}$$

By integration and rearrangement of Eq. (20)

$$kX = -4\pi(D_X + D_W)\gamma\int_{X=X_R}^{X} d\{X_r \exp\left(\phi(r)/k_BT\right)\} \tag{21}$$

and here

$$\gamma^{-1} = \int_R^\infty \frac{\exp\left(\phi(r)/k_BT\right)}{r^2}dr \tag{22}$$

Because $r\text{-}>\infty$, $V(r)$ approaches zero

$$\begin{aligned} k &= \frac{4\pi(D_X + D_W)\gamma}{X}\{X - X_R\exp\left(\phi(r)/k_BT\right)\} \\ &= \frac{4\pi(D_X + D_W)\gamma}{1 + \left[4\pi(D_X + D_W)\gamma/k_R\exp\left(-\phi(r)/k_BT\right)\right]} \\ &\approx \frac{4\pi D_X\gamma}{1 + \left[4\pi D_X\gamma/k_R\exp\left(-\phi(r)/k_BT\right)\right]} \end{aligned} \tag{23}$$

Here, $X_{r=R} = (k/k_R)\, X$. In calculating the above formula, fluctuations of diffusion coefficients are neglected when the concentration of X is sufficiently low and kept constant during the reaction. The kinetic coefficient is controlled by the diffusion process and the aggregation or assembly is completed through interactions between X and W; $X_{r=R}$ is set to zero. Then, k is given by Eq. (23):

$$k = 4\pi D_X R \tag{24}$$

The diffusion coefficients can be altered in proportion to the fluctuation of monomeric protein concentration [9–14]. In the derivation of Eq. (23), the diffusion coefficient is related to the viscosity η by the Einstein-Stokes formula

$$D = \frac{k_B T}{6\pi r \eta(T)} \tag{25}$$

By using the Gibbs-Duhem expression, the diffusion coefficient D of one macromolecule can be written as

$$D = \frac{k_B T}{\eta}\left(1 - \frac{N_A v_1}{M_1} c_1\right)(1 + 2A_1 M_1 c_1 + \cdots) \triangleq D_0\left(1 - \frac{N_A v_1}{M_1} c_1\right)(1 + 2A_1 M_1 c_1 + \cdots) \tag{26}$$

where T is the temperature of the solution, k_B is the Boltzmann constant, and η_1 is the frictional coefficient of the macromolecule in solution. A_1 is the second virial coefficient, v_1 is the partial specific volume of protein with molecular weight M_1, and N_A is Avogadro's number. The small letter c_1 denotes the concentration of the solute macromolecule. Then, dependency of the diffusion coefficient on the ith component, D_i, is as follows from (26):

$$a_{ij} \equiv \frac{\partial D_i}{\partial c_j} = D_{0i}\left(2A_j M_j - \frac{N_A v_j}{M_j}\right) \tag{27}$$

where v_j is the partial specific volume of the polymer with molecular weight M_j.

Further, the diffusion coefficient is given by extending the above formula to a mixed solution of two macromolecules, X and Z

$$\begin{aligned} D_X(X,Z) &= \frac{k_B T}{\eta_X}\left(1 - \frac{N_A v_X}{M_X} X - \frac{N_A v_Z}{M_Z} Z\right)(1 + 2A_X M_X X + 2A_Z M_Z Z + \cdots) \\ D_Z(X,Z) &= \frac{k_B T}{\eta_Z}\left(1 - \frac{N_A v_X}{M_X} X - \frac{N_A v_Z}{M_Z} Z\right)(1 + 2A_X M_X X + 2A_Z M_Z Z + \cdots) \end{aligned} \tag{28}$$

where v_X and v_Z are the partial specific volumes of X and Z with molecular weights M_X and M_Z, respectively. A_X and A_Z are the second virial coefficients.

2.5. Fluctuation of diffusion coefficient

Subsequently, let us consider the fluctuation of participant proteins using lowercase letters

$$X = X_e + x \tag{29}$$

$$Z = Z_e + z \tag{30}$$

Here, the sum of the kinetic rate of fluctuations is constant because total amount of monomeric protein is constant

$$\dot{x} + \dot{y} + \dot{z} = 0 \tag{31}$$

The dependency of the diffusion rate refers to the high interaction activities of X and low interaction activities of Z. An increase in X contributes to a decrease in the diffusion coefficients D_X and D_Z in fluctuation items because of the higher interaction activity that reduces diffusion; in contrast, an increased Z contributes to increased diffusion coefficients because the interaction between increased Z induces lower assembly activity for the interaction with monomeric proteins, resulting in increased mobility of monomeric proteins. This dependency gives the non-linearity fluctuation items in the kinetic equation. Here, all preparations for kinetics are completed.

2.6. General theory of non-linear kinetic equation of protein assembly

According to the above simple reaction cascade, (1)-(5), the kinetic equation contains the concentrations of monomeric proteins as variables. For simplification, the equations are written as follows:

$$\dot{X} = -k_1 D_1 WX + k_3 PZ - k_4 D_4 X^2 - k_5 D_5 XZ \tag{32}$$

$$\dot{W} = k_1 D_1 WX - k_2 D_2 W \approx 0 \tag{33}$$

$$\dot{Z} = k_2 D_2 W - k_3 PZ + k_4 D_4 X^2 + k_5 D_5 XZ \tag{34}$$

Here,

$$D_1 \triangleq \frac{D_X + D_W}{2} \sim \frac{D_X}{2}, D_4 \triangleq D_X, D_5 \triangleq \frac{D_X + D_Z}{2} \tag{35}$$

Further, for simplicity, Eqs. (34) and (36) are given by replacing the kinetic coefficients with arbitrary coefficients

$$\dot{X} = -k_1{}' WX + k_3 PZ - k_4{}' X^2 - k_5{}' XZ \tag{36}$$

$$\dot{Z} = k_2{}' W - k_3 PZ + k_4{}' X^2 + k_5{}' XZ \tag{37}$$

Here, $k_1 D_1 W = k_1'$, $k_4 D_4 W = k_4'$, $k_5 D_5 = k_5'$, and $p = k_3 P$. At the steady state, setting the right hands of Eqs. (32)-(34) equal to zero

$$X_e = \frac{k_2{}'}{k_1{}'}, Z_e = \frac{k_2{}'\left(k_{4_4}'k_2{}' + k_1{}^{2'} W\right)}{k_1{}'(k_1{}'P - k_2{}'k_5{}')} \tag{38}$$

Here, the fluctuation of diffusion coefficients are given by

$$a = \frac{\partial k_1'}{\partial x}, b = \frac{\partial k_1'}{\partial z}, c = \frac{\partial k_4'}{\partial x}, d = \frac{\partial k_4'}{\partial z}, e = \frac{\partial k_5'}{\partial x}, f = \frac{\partial k_5'}{\partial z} \tag{39}$$

By altering X, Z, k_1', k_4', and k_5' into $X + x$, $Z + z$, $k_1' - ax + bz$, $k_4' - cx + dz$, and $k_5' - ex + fz$ in Eqs. (36) and (37) and arranging, two dependent equations are obtained

$$\dot{x} = f_x x + (f_z + p)z + f_{xx}x^2 + f_{xz}xz + f_{zz}z^2 \tag{40}$$

$$\dot{z} = f_x' x + \left(f_z' - f_z(p)\right)z + f_{xx}'x^2 + f_{zx}'xz + f_{zz}'z^2 \tag{41}$$

Accordingly, the overall behavior of the kinetics of protein assembly is given by the monomeric kinetics of x and z. $fxz^{(')}$, $fzx^{(')}$, $fxx^{(')}$, and $fzz^{(')}$ represent the assembly activity between X and Z, Z and X, X themselves, and Z themselves.

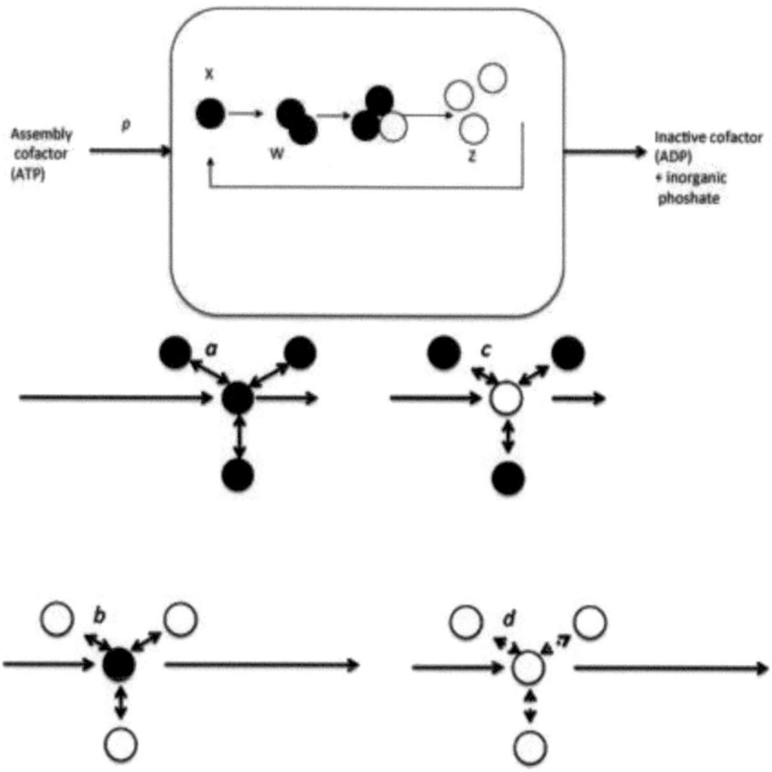

Figure 2. Scheme of monomer interaction. Individual globules represent monomers X•, and Z ○, released species. The supply of the cofactor is kept constant and inactive cofactor is released continuously. The differential coefficients *a*, *b*, *c*, and *d* indicate the interaction activity between X and Z. The differential coefficients are given in Eq. (39). The interaction activity between X is higher and therefore the diffusion rate of X between X becomes slower; the interaction activity between X and Z is lower, and therefore the diffusion rate *of X/Z* between Z/X becomes slightly slower; the interaction activity between Z and Z is negligible and therefore the change in the diffusion rate of Z through Z is negligible.

In reality,

$$\begin{aligned}\dot{x} &= -\{W(k_1{}' - aX_e) + 2k_4{}'X_e + k_5{}'Z_e\}x + (Wa - k_4{}' + 2cX_e + eZ_e)x^2 \\ &+ (p - bX_e - k_5{}'X_e - dX_e{}^2 - fX_eZ_e)z - (k_5{}' + Wb - eX_e + f_eZ_e)xz - fX_ez^2\end{aligned} \tag{42}$$

$$\begin{aligned}\dot{z} &= (2k_4{}'X_e + k_5{}'Z_e - cX_e{}^2 - eX_eZ_e)x + (k_4{}' - 2cX_e - eZ_e)x^2 \\ &+ (-p + k_5{}'X_e + dX_e{}^2 + fX_eZ_e)z + (k_5{}' + 2dX_e - eX_e + f_eZ_e)xz + fX_ez^2\end{aligned} \tag{43}$$

Thus, we determined a general formula for protein assembly. Cytoskeletal protein and protein complexes in the signaling cascade can be described using this formula.

2.7. Linearity of cofactor supply to the assembly system

While protein assembly is a non-linear reaction involving a complicated set of reactions or assembly steps, the supply of cofactor is simply given by the linear kinetic rate items as shown in Eqs. (36, 37, 40 and 41). This means that the supply rate will be essentially be an order parameter of the assembly system and is controllable by altering the concentration of the cofactor (**Figure 2**). Accordingly, parameter p is variable in the numerical simulation, as described below.

3. Calculus simulation of concentration oscillation

3.1. Oscillation of monomer concentration fluctuation

In actual simulation of protein assembly, numerical calculation was performed over a sufficiently long period to evaluate the trend in system behavior.

Simulation: A simulation was performed using Mathematica® version 8 (Wolfram Research, Champaign, IL, USA).

Simulation was performed with the notation in Eqs. (42) and (43), in which p is (a) 0.8, (b) 0.81, and (c) 1.00 (**Figure 3**).

Below is the simulation program cord using Mathematica ver. 8 when p = 0.8:

D1 = 0.27, k2 = 0.00035, a = 790, b = 650, c = 105, d = 105, e = 105, f = 105, p = 0.8, Dxx = 155, Dxz = 155, W = 1.

X = k2/D1.

Z = (k2 (D1^2 W + Dxx k2))/(D1 (D1 p - Dxz k2)).

NDSolve[{x′[t] == − (W (D1 - a X) + 2 X Dxx + Dxz Z) x[t] + (W a - Dxx + 2 c X + e Z) x[t]^2 + (p - Dxz X - b X - d X^2 - f X Z) z[t] - (Dxz + W b - e X + f Z) x[t] z[t] - (f X) z[t]^2,

z′[t] == (2 X Dxx + Dxz Z - c X^2 - e X Z) x[t] + (Dxx - 2 c X - e Z) x[t]^2 + (Dxz + 2 X d - e X + f Z) x[t] z[t] + (Dxz X - p + d X^2 + f X Z) z[t], x[0] == 0.000001, z[0] == 0.000001}, {x, z}, {t, 0, 30,000}, MaxSteps - > 50,000].

http://dx.doi.org/10.5772/intechopen.70750

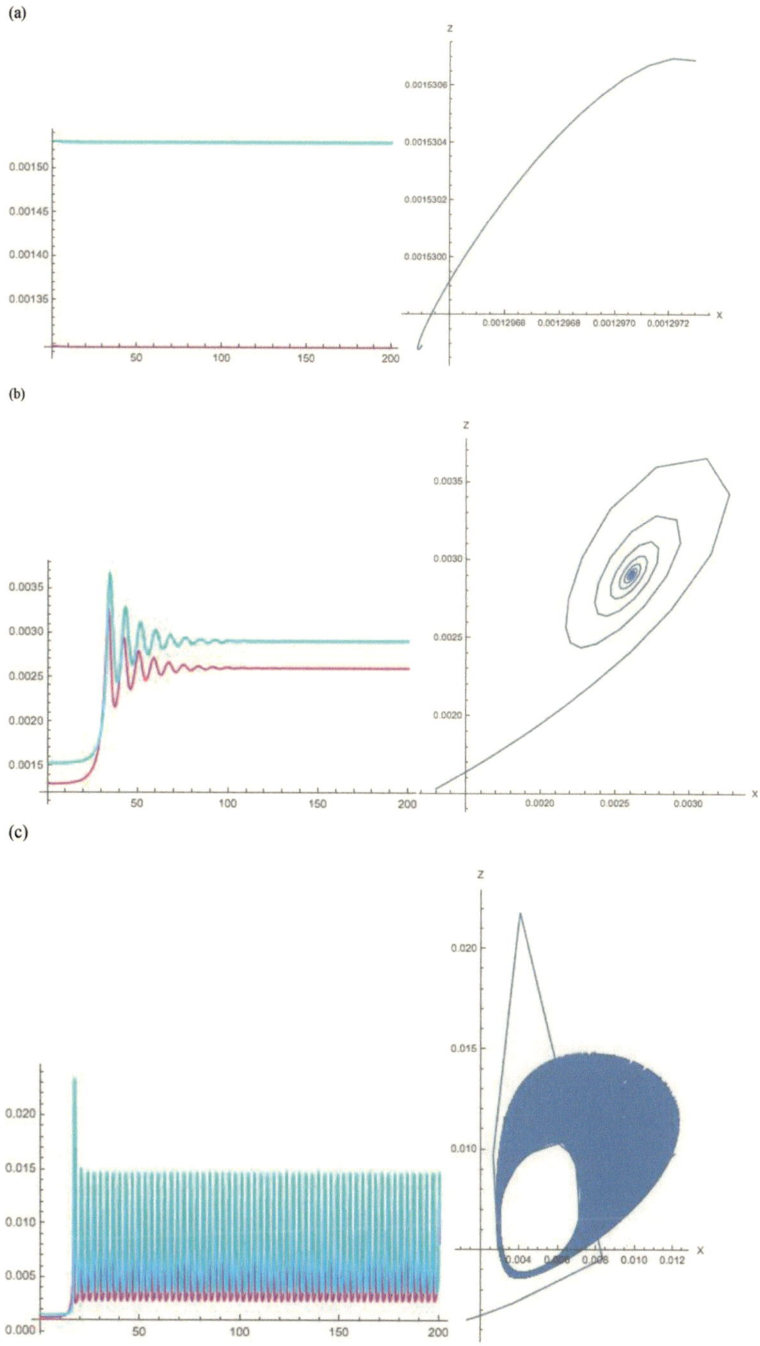

Figure 3. Numerical simulation of protein assembly. Diffusion of active cofactor binding signaling molecule (X) and inactive cofactor binding signaling molecule (Z). The upper graph shows two parametric plots of X and Z. Red and blue lines in the lower graph represent the concentrations of X and Z, respectively. The horizontal axis represents time (s) ($0 \le t \le 200$) and vertical axis represents the concentrations of X and Z, respectively. When p exceeds 0.80, chaos-like oscillation is observed. Mathematica version 8 was used.

g001 = Plot[{X + x[t]} /. %, {t, 0, 200}, PlotRange - > All, PlotStyle - > {RGBColor[1, 0, 1]}, PlotRange - > ALL].

g002 = Plot[{Z + z[t]} /. %%, {t, 0, 200}, PlotRange - > All, PlotStyle - > {RGBColor[0, 1, 1]}, PlotRange - > All].

g003 = ParametricPlot[Evaluate[{X + x[t], Z + z[t]} /. %%%], {t, 0, 2000}, PlotRange - > All, AxesLabel - > {"X", "Z"}] Show[g001, g003, AxesLabel - > {"t", "X, Z"}].

As a result, regular oscillation with fluctuations in the amplitude and frequency can be illustrated in the plot following the above calculation. Plots on the right side that the oscillation becomes definite within a limit-cycle when p = 1.5.

3.2. Oscillation frequency

In a previous study [6], an interesting relationship was observed between the average frequency of simulated oscillation of the monomeric proteins and difference $p - p_c$. The frequency was nearly equivalent, but with irregular fluctuations, except for during the initial phase. The relationship between average frequency < f > and $p - p_c$ is given by:

$$< f > \simeq 0.0256 \ln \left(p - p_c\right) + 0.1407 \quad (44)$$

These formulae imply that the amplitude of monomeric protein fluctuation provides information regarding the outside alteration of the cofactor. Thus, outside alteration is transformed inside into the information of assembly.

4. Center manifold theory (CMT)

4.1. Center manifold formulae

For stability analysis of the nonlinear dynamics in protein assembly, the center manifold theory (CMT) for non-linear dynamic biological systems has been applied. Simulation is oriented to analyze the behavior around critical values of the order parameter. CMT has been applied to the Lotka-Volterra model of the predator-prey system to provide important simulation results [33, 34]. However, the CMT can be applied to the protein assembly model. For stability analysis around the critical point, Eqs. (40) and (41) were formulated. When p is equivalent to p_c, the Jacobian matrix L for (x, z) is given using the linear coefficients of (x, z) in Eqs. (40) and (41):

$$L = \begin{pmatrix} f_x & f_z + f(p_c) \\ f_x' & f_z' - F(p_c) \end{pmatrix} \quad (45)$$

Particularly, the function $f(p)$ represents the input and output of the cofactor that is the order parameter. Using the eigenvectors of L, $(\mathbf{l}_1, \mathbf{l}_2)$, coordinate transformation into u and v is performed as follows:

$$\frac{d}{dt}\begin{bmatrix} x \\ z \end{bmatrix} = \frac{d}{dt}[\mathbf{l}_1 \ \ \mathbf{l}_2]\begin{bmatrix} u \\ v \end{bmatrix} \tag{46}$$

$$\frac{d}{dt}\begin{bmatrix} u \\ v \end{bmatrix} = \frac{d}{dt}[\mathbf{l}_1 \ \ \mathbf{l}_2]^{-1}\begin{bmatrix} x \\ z \end{bmatrix} \tag{47}$$

The above formulae are subsequently set as

$$du/dt = f_u(u, v) \tag{48}$$

$$dv/dt = f_v(u, v) \tag{49}$$

The center manifold around the critical point ($p = p_c$) is then given as follows:

$$u = h(\varepsilon, v) = a_1 v^2 + a_2 v\varepsilon + a_3\varepsilon^2 + a_4 v^3 + a_5 v^2\varepsilon + a_6 v\varepsilon^2 + a_7\varepsilon^3 + O(\varepsilon^4) \tag{50}$$

The effect of changing p and ε ($p = p_c$) is analyzed using the center manifold around the critical point of the system. Subsequently

$$u = (dv/dt)\partial h(u, \varepsilon)/\partial u + (d\varepsilon/dt)\partial h(u, \varepsilon)/\partial\varepsilon = (2a_1 v + a_2\varepsilon)f_u(u, v) \tag{51}$$

Using Eqs. (49) and (50)

$$(2a_1 v + a_2\varepsilon)f_u(u, v) = a_1 v^2 + a_2 v\varepsilon + a_3\varepsilon^2 + a_4 v^3 + a_5 v^2\varepsilon + a_6 v\varepsilon^2 + a_7\varepsilon^3 + O(\varepsilon^4) \tag{52}$$

Solving Eq. (52) gives the coefficients of a_i in Eq. (50): $a_3 = a_7 = 0$. Substituting u in Eq. (51) given by v and ε into $f_v(u, v)$ in Eq. (47), the kinetic stability equation is given for fluctuation v using the coefficients n_i ($i = 1, \ldots, 7$) as follows:

$$dv/dt = n_1 v^2 + n_2 v\varepsilon + n_3\varepsilon^2 + n_4 v^3 + n_5 v^2\varepsilon + n_6 v\varepsilon^2 + n_7\varepsilon^3 + O(\varepsilon^4) \tag{53}$$

Independently of the numerical values in Eq. (53)

$$n_3, n_6, n_7 = 0 \tag{54}$$

Using this result, we have

$$dv/dt = n_1 v^2 + n_2 v\varepsilon + n_4 v^3 + n_5 v^2\varepsilon + O(\varepsilon^4) \tag{55}$$

By setting the left-hand side of Eq. (55) equivalent to zero, a Hopf-bifurcation of the given system is shown

$$v = 0, \quad \frac{-n_1 - n_5\varepsilon \pm \sqrt{(n_1 + n_5\varepsilon)^2 - 4n_2 n_4\varepsilon}}{2n_4} \tag{56}$$

Further approximate solutions to Eq. (56) are given as

$$v = 0, \approx c\varepsilon, \; - n_1/n_4 \tag{57}$$

Thereafter, the solution $c\varepsilon$ increases as the concentration of ATP/GTP increases. From Eq. (51), u is formulated using an arbitrary coefficient c

$$u \approx 0, c(n_1/n_4)^2 \tag{58}$$

This result implies that the fluctuation has two different values for the amplitude of an oscillation.

5. Prospects of protein analysis assembly

Our aim in this review was to evaluate the association between the non-linearity in protein diffusion and assembly. Here, we reviewed studies of protein assembly or interactions [1–4] and performed mathematical analysis of the model in addition to numerical simulations. Because the assumptions of the model are minimal, the simulation provides insight into assembly. The results are summarized as follows: (i) the non-linear kinetic equations including only two independent parameters may reveal dynamic behavior in the fluctuation of the monomer concentration, (ii) the increase near the critical concentration of the cofactor induces oscillations in amplitude and frequency; and (iii) center manifold analysis predicts the stability of the model system near the critical concentration, showing bifurcation with respect to the cofactor supply value. The behavior of the system shown in the simulation indicates that the concentration change information of a cofactor outside the system is transduced into another type of information, e.g., frequency of the concentration oscillation of the monomer. A small increase in the outside cofactor concentration induces an oscillation change inside the monomeric protein, which may be crucial for responding to transformations in the outside environment. Such a trajectory in the observed oscillation resembles a limit-cycle-like in the well-known two-parametric Lorenz model [13, 14].

Previous systems biology models did not focus on the diffusion process of a protein in the cytoplasm. Non-linearity in the process is critical and essential to protein assembly. Before considering a set of simultaneous kinetic equations, non-linearity in the diffusion process should be considered, as *de novo* nucleation is negligible compared with the reaction of the monomer and oligomer. The interaction between assembly-active monomer proteins attenuates the diffusion rate in a non-linear fashion because they can assemble, which inevitably yields non-linearity. As shown in this review, CMT is useful for reducing the parameters of detailed stability analysis around the critical state.

6. Conclusion

Protein interactions play an important role in various biological activities at the cell level. Although protein diffusion is a rate-limiting-step, cell behavior is orchestrated by protein interactions, and signaling transduction, the cytoskeleton, and cell motility are dynamically altered in processes similar to "phase transitions" in inorganic chemical reactions. This review provides a model of phase transition affected by minimal changes in cofactor concentration; but oscillation and bifurcation are inducible by the simple model. Systems biology multi-parametric analysis remains important; however, a simple model sufficiently can better illustrate oscillations in protein concentration with a limit-cycle. Outside alteration such as cofactor concentration change is transformed inside into the information of assembly. Stability analysis using the CMT is a simple method for understanding protein-interacting systems.

Acknowledgements

This work was supported by the Ministry of Education, Culture, Sport, Science and Technology, Japan, under the project name "Synergy of Fluctuation and Structure." Project number 2502. The funders had no role in the study design, data collection, and analysis; decision to publish; or preparation of the manuscript.

Conflicts of interest

The authors have declared no conflicts of interest.

Author details

Tatsuaki Tsuruyama

Address all correspondence to: tsuruyam@kuhp.kyoto-u.ac.jp

Department of Pathology, Drug Discovery Medicine, Graduate School of Medicine, Kyoto University, Kyoto, Kyoto Prefecture, Japan

References

[1] Gorecki J, Gorecka JN, Nowakowski B, Ueno H, Tsuruyama T, Yoshikawa K. Sensing parameters of a time dependent inflow with an enzymatic reaction. In: Adamatzky A, editor. Advances in Unconventional Computing, Emergence, Complexity and Computation. Vol. 23; 2017. p. 85-104

[2] Tsuruyama T. A model of cell biological signaling predicts a phase transition of signaling and provides mathematical formulae. PloS One. 2014;**9**(7):e102911

[3] Tsuruyama T. Kinetic stability analysis of protein assembly on the center manifold around the critical point. BMC Systems Biology. 2017;**11**(1):13

[4] Ueno H, Tsuruyama T, Nowakowski B, Gorecki J, Yoshikawa K. Discrimination of time-dependent inflow properties with a cooperative dynamical system. Chaos. 2015;**25**(10):103115

[5] Di Camillo B, Carlon A, Eduati F, Toffolo GM. A rule-based model of insulin signalling pathway. BMC Systems Biology. 2016;**10**(1):38

[6] Babu CVS, Song EJ, Yoo YS. Modeling and simulation in signal transduction pathways: A systems biology approach. Biochimie. 2006;**88**(3–4):277-283

[7] Selimkhanov J, Taylor B, Yao J, Pilko A, Albeck J, Hoffmann A, Tsimring L, Wollman R. Systems biology. Accurate information transmission through dynamic biochemical signaling networks. Science. 2014;**346**(6215):1370-1373

[8] Zhao M, Kong L, Qu H. A systems biology approach to identify intelligence quotient score-related genomic regions, and pathways relevant to potential therapeutic treatments. Scientific Reports. 2014;**4**:4176

[9] Hazra P, Inoue K, Laan W, Hellingwerf KJ, Terazima M. Tetramer formation kinetics in the signaling state of AppA monitored by time-resolved diffusion. Biophysical Journal. 2006;**91**(2):654-661

[10] Wu Z, Wang HW, Mu W, Ouyang Z, Nogales E, Xing J. Simulations of tubulin sheet polymers as possible structural intermediates in microtubule assembly. PloS One. 2009;**4** (10):e7291

[11] VanBuren V, Cassimeris L, Odde DJ. Mechanochemical model of microtubule structure and self-assembly kinetics. Biophysical Journal. 2005;**89**(5):2911-2926

[12] Hamon L, Savarin P, Curmi PA, Pastré D. Rapid assembly and collective behavior of microtubule bundles in the presence of polyamines. Biophysical Journal. 2011;**101**(1):205-216

[13] Wattis JAD, Coveney PV. Mesoscopic models of nucleation and growth processes: A challenge to experiment. Physical Chemistry Chemical Physics. 1999;**1**:2163-2176

[14] Carlsson AE. Model of reduction of actin polymerization forces by ATP hydrolysis. Physical Biology. 2008;**5**(3):036002

[15] Brooks FJ, Carlsson AE. Actin polymerization overshoots and ATP hydrolysis as assayed by pyrene fluorescence. Biophysical Journal. 2008;**95**(3):1050-1062

[16] Oosawa F, Kasai M. A theory of linear and helical aggregations of macromolecules. Journal of Molecular Biology. 1962;**4**:10-21

[17] Michaels TC, Garcia GA, Knowles TP. Asymptotic solutions of the Oosawa model for the length distribution of biofilaments. The Journal of Chemical Physics. 2014;**140**(19):194906

[18] Chretien D, Jainosi I, Taveau JC, Flyvbjerg H. Microtubule's conformational cap. Cell Structure and Function. 1999;**24**(5):299-303

[19] Zilberman M, Sofer M. A mathematical model for predicting controlled release of bioactive agents from composite fiber structures. Journal of Biomedical Materials Research. Part A. 2007;**80**(3):679-686

[20] Oosawa F, Asakura S. Thermodynamics of the Polymerisation of Proteins; New York: Academic Press, 1975. p. 204

[21] Zapperi S, Mahadevan L. Dynamic instability of a growing adsorbed polymorphic filament. Biophysical Journal. 2011;**101**(2):267-275

[22] Hammele M, Zimmermann W. Modeling oscillatory microtubule polymerization. Physical Review. E, Statistical, Nonlinear, and Soft Matter Physics. 2003;**67**(2 Pt 1):021903

[23] Chen BS, CC W. On the calculation of signal transduction ability of signaling transduction pathways in intracellular communication: Systematic approach. Bioinformatics. 2012;**28**(12):1604-1611

[24] Zumsande M, Gross T. Bifurcations and chaos in the MAPK signaling cascade. Journal of Theoretical Biology. 2010;**265**(3):481-491

[25] Arnal I, Karsenti E, Hyman AA. Structural transitions at microtubule ends correlate with their dynamic properties in Xenopus egg extracts. Journal of Cell Biology. 2000;**149**(4): 767-774

[26] Daga RR, Lee KG, Bratman S, Salas-Pino S, Chang F. Self-organization of microtubule bundles in anucleate fission yeast cells. Nature Cell Biology. 2006;**8**(10):1108-1113

[27] Kasas S, Cibert C, Kis A, De Los Rios P, Riederer BM, Forro L, Dietler G, Catsicas S. Oscillation modes of microtubules. Biology of the Cell. 2004;**96**(9):697-700

[28] Bauer KC, Göbel M, Schwab ML, Schermeyer MT, Hubbuch J. Concentration-dependent changes in apparent diffusion coefficients as indicator for colloidal stability of protein solutions. International Journal of Pharmaceutics. 2016;**511**(1):276-287

[29] Gallagher WH, Woodward CK. The concentration dependence of the diffusion coefficient for bovine pancreatic trypsin inhibitor: A dynamic light scattering study of a small protein. Biopolymers. 1989;**28**(11):2001-2024

[30] Giavazzi F, Fornasieri A, Vailati A, Cerbino R. Equilibrium and non-equilibrium concentration fluctuations in a critical binary mixture. The European Physical Journal. E, Soft Matter. 2016;**39**(10):103

[31] Giavazzi F, Savorana G, Vailati A, Cerbino R. Structure and dynamics of concentration fluctuations in a non-equilibrium dense colloidal suspension. Soft Matter. 2016;**12**(31): 6588-6600

[32] Chang X, Wei J. Stability and Hopf bifurcation in a diffusive predator-prey system incorporating a prey refuge. Mathematical Biosciences and Engineering. 2013;**10**(4):979-996

[33] Zhang X, Zhao H. Bifurcation and optimal harvesting of a diffusive predator-prey system with delays and interval biological parameters. Journal of Theoretical Biology. 2014;**363**: 390-403

[34] Xiao M, Zheng WX, Cao J. Hopf bifurcation of an (n + 1)-neuron bidirectional associative memory neural network model with delays. IEEE Transactions on Neural Networks and Learning Systems. 2013;**24**(1):118-132

Chapter 7

Plasma Kinetic Theory

Kashif Chaudhary, Auwal Mustapha Imam,
Syed Zuhaib Haider Rizvi and Jalil Ali

Additional information is available at the end of the chapter

http://dx.doi.org/10.5772/intechopen.70843

Abstract

The description of plasma using fluid model is mostly insufficient and requires the consideration of velocity distribution which leads to kinetic theory. Kinetic theory of plasma describes and predicts the condition of plasma from microscopic interactions and motions of its constituents. It provides an essential basis for an introductory course on plasma physics as well as for advanced kinetic theory. Plasma kinetics deals with the relationship between velocity and forces and the study of continua in velocity space. Plasma kinetics mathematical equations provide aid to the readers in understanding simple tools to determine the plasma dynamics and kinetics as described in this chapter. Kinetic theory provides the basics and essential introduction to plasma physics and subsequently advanced kinetic theory. Plasma waves, oscillations, frequencies, and applications are the subjects of kinetic theory. In this chapter, mathematical formulations essential for exploring plasma kinetics are compiled and described simplistically along with a precise discussion on basic plasma parameters in simple language with illustrations in some cases.

Keywords: plasma parameters, kinetic theory, particle distribution

1. Introduction

Plasma is the fourth state of matter, and it is defined as "a quasineutral gas of charged and neutral particles which exhibits collective behavior." As plasma contains charged particles, these charged particles move around and generate local concentrations of positive or negative charges (collective behavior) which give rise to electric fields. Motion of these charges also generates currents and hence magnetic fields [1]. Therefore, the macroscopic forces acting in plasma are totally different from ordinary gasses and hence remarkable differences in their physical properties are observed. The salient features of the plasma can be understood by investigating the behavior of the electrons, by far the most mobile-charged particle in plasma [2].

Plasma physics deals with the equilibrium and non-equilibrium properties of a statistical system of charged particles. Microscopic degrees of freedom arising from the motion of individual particles describe the system. These statistics therefore theoretically treat the macroscopic behavior of such a system [3].

The knowledge of plasma parameters helps to understand the dynamics of plasma. Electrons being dominant mobile species play an important role in the behavior of the plasma. The most important of these parameters include plasma temperature, electron density, Debye shielding, and Debye length. Plasma is transient in nature. Therefore, the plasma is generally characterized on the basis of instantaneous observations. Charged particles, neutrals, and molecules coexist in plasma under various circumstances. Conditions in the plasma strongly depend on the distribution of charged particles, where electrons being lighter and highly mobile play a dominant role. Therefore, plasma is generally represented through parameters which are derived from the behavior of electrons in the plasma. When an external point charge is introduced or a localized unbalanced charge is formed in the plasma, readjustment of charge density occurs to neutralize the effect by shielding its electric field. The electrons being more mobile than heavier ions move toward or away from the unbalanced charge faster than ions. It gives rise to oscillations which are referred to as *electron oscillations* or *plasma oscillations*. The frequency at which these oscillations take place is called as *plasma frequency* [4]. This phenomenon of shielding or screening a foreign charge or an unbalanced charge inside the plasma is known as *Debye shielding* or sometimes referred to as *Debye screening*. It is specific for plasma. Charges keep accumulating around the foreign or unbalanced charge until the static electric field of the unbalanced charge is screened and the balance is restored. A sphere of charges that is created around the unbalanced charge is called as Debye sphere and its radius is known as *Debye length*. A detailed discussion on two of the most important plasma parameters *electron density* and *plasma temperature* is provided in Sections 2.7 and 2.8.

Comprehensive mathematical details on certain aspects of plasma are found in published literature. The authors aim to provide readers with an overall view of fundamental mathematical relations explaining the kinetics of plasma that are compiled simplistically in one chapter. This can be an easy reference for the researchers interested in plasma kinetics.

2. Plasma kinetic equations

Plasma physics involves phenomena that are related to dynamical processes in statistical mechanics. It is thus very significant to study the properties and structure of the basic kinetic equations governing the dynamical behavior of plasma [5]. The dynamical behavior of a system of N-interacting particles is generally investigated using the Liouville equation. A microscopic distribution function could be used to describe the behavior of such a system. A six-dimensional phase-space distribution function called "Klimontovich distribution function," which obeys a continuity equation in the phase space, is defined. The system of charged particles can then be described by Bogoliubov-Born-Green-Kirkwood-Yvon (BBGKY) hierarchy equations.

http://dx.doi.org/10.5772/intechopen.70843

2.1. Klimontovich equation

To further formalize the kinetic theory, we introduce Klimontovich's microscopic description and derivation of the Bogoliubov-Born-Green-Kirkwood-Yvon equations [6]. To introduce the Klimontovich equation, we consider a classical system containing N identical particles in a box of volume V; $n \equiv N/V$ denoting the average number density. Each particle in the box is characterized by an electric charge q and mass m.

In the six-dimensional phase space consisting of the position $\boldsymbol{r}$ and velocity $\boldsymbol{v}$, each of the particle has its own trajectory; for ith particle,

$$\mathrm{Xi(t)} \equiv [\boldsymbol{r}_i(\mathrm{t}), \boldsymbol{v}_i(\mathrm{t})]. \tag{1}$$

The microscopic density of the particles in the phase space may be expressed by the summation of the six-dimensional delta functions as

$$N(X;t) \equiv \left(\frac{1}{n}\right) \sum_{i=1}^{n} \delta[X - X_i(t)] \tag{2}$$

where $X \equiv (\boldsymbol{r}, \boldsymbol{v})$. $N(\boldsymbol{X}; t)$ is known as the *Klimontovich distribution function* which satisfies the continuity equation in the phase space,

$$\frac{dN}{dt} = \frac{\partial N}{\partial t} + \dot{\mathrm{X}}.\frac{\partial N}{\partial \mathrm{X}} = 0. \tag{3}$$

In phase-space coordinates, Eq. (3) can be written as

$$\frac{dN}{dt} + \boldsymbol{v}.\frac{\partial N}{\partial \boldsymbol{r}} + \dot{\boldsymbol{v}}.\frac{\partial N}{\partial \boldsymbol{v}} = 0 \tag{4}$$

where $\dot{\boldsymbol{v}}$ is the acceleration at point $(\mathbf{r}, \mathbf{v})$.

The electromagnetic acceleration is very important in plasma physics,

$$\dot{\boldsymbol{v}} = \frac{q}{m}\left[\boldsymbol{E}(\boldsymbol{r}, t) + \frac{v}{c} \times \boldsymbol{B}(\boldsymbol{r}, t)\right]. \tag{5}$$

The electric and magnetic fields $\boldsymbol{E}(\boldsymbol{r}, t)$ and $\boldsymbol{B}(\boldsymbol{r}, t)$ consist of two separate contributions: those applied from the external sources and those produced from the microscopic fine-grained distribution of the charged particles.

$$\boldsymbol{E}(\boldsymbol{r}, t) = \boldsymbol{E}_{ext}(\boldsymbol{r}, t) + \boldsymbol{e}(\boldsymbol{r}, t), \boldsymbol{B}(\boldsymbol{r}, t) = \boldsymbol{B}_{ext}(\boldsymbol{r}, t) + \boldsymbol{b}(\boldsymbol{r}, t). \tag{6}$$

The microscopic fine-grained fields $\boldsymbol{e}(\boldsymbol{r}, t)$ and $\boldsymbol{b}(\boldsymbol{r}, t)$ are determined from a solution of Maxwell equations,

$$\nabla \times e + \frac{1}{c}\frac{\partial \boldsymbol{b}}{\partial \boldsymbol{t}} = 0,\ \nabla \times b - \frac{1}{c}\frac{\partial \boldsymbol{e}}{\partial \boldsymbol{t}} = \frac{4\pi}{\mathrm{c}} qn \int \boldsymbol{v} N(X;t) d\boldsymbol{v},\ \nabla.\boldsymbol{e} = 4\pi qn \left[\left[\int N(X;t) d\boldsymbol{v} - 1\right],\ \nabla.\boldsymbol{b} = 0. \tag{7}$$

For a given $N(X;t)$, the solution to these set of equations can generally be written; the solution, when substituted in Eq. (5), would amount to taking account of both electromagnetic and electrostatic interactions between the particles. The electromagnetic interactions are usually negligible as compared with the electrostatic interactions for a nonrelativistic plasma; hence, the microscopic fields become.

$$\boldsymbol{e}(\boldsymbol{r},t) = -qn\frac{\partial}{\partial \boldsymbol{r}}\int\frac{\mathbf{N}(\mathbf{X}';\boldsymbol{t})}{|\boldsymbol{r}-\mathbf{r}'|}d\mathbf{X}', \boldsymbol{b}(\boldsymbol{r},t) = \mathbf{0}. \tag{8}$$

Substituting Eq. (8) into Eq. (6), we get an expression for the acceleration in terms of $N(X;t)$. Eq. (3) can be written with the aid of such an expression as

$$\left[\left[\frac{\partial}{\partial t} + \boldsymbol{L}(\mathrm{X}) - \int \mathbf{V}(\mathrm{X},\mathrm{X}')\mathrm{N}(\mathrm{X}';\mathrm{t})d\mathrm{X}'\right]N(X;t) = 0 \tag{9}$$

$\boldsymbol{L}(X)$, as a single particle operator is defined by

$$\boldsymbol{L}(X) \equiv \boldsymbol{v}.\frac{\partial}{\partial \boldsymbol{r}} + \frac{q}{m}\left[\boldsymbol{E}_{ext}(\boldsymbol{r},t) + \frac{v}{c} \times \boldsymbol{B}_{ext}(\boldsymbol{r},t)\right].\frac{\partial}{\partial \boldsymbol{v}} \tag{10}$$

And $V(X,X')$ is a two-particle operator arising from the Coulomb interaction defined by

$$V(X,X') \equiv \frac{q^2 n}{m}\left[\frac{\partial}{\partial \boldsymbol{r}}\frac{1}{|\boldsymbol{r}-\boldsymbol{r}'|}\right].\frac{\partial}{\partial v} \tag{11}$$

Eq. (9) is known as *Klimontovich equation*. The equation describes the space–time evolution of the microscopic distribution function.

2.2. Liouville distribution

The fine-grained distribution function, which is precise in describing the microscopic conditions of many particles, would not by itself correspond to the coarse-grained quantities in the macroscopic view. There is a need to introduce an averaging process based on the Liouville distribution over the 6 N–dimensional phase space to establish a connection between them.

In T-space, the microscopic state of the system is expressed by a point.

$\{X_i\} \equiv (X_1, X_2, \ldots\ldots, X_n)$, called a system point.

By following a normal procedure of the ensemble theory in statistical mechanics, it can assume N replicas which are microscopically identical to the system under consideration. N can be chosen to be very large so that it can approach infinity when it requires. These N replicas are characterized by different microscopic configurations; the system points are scattered over the T-space. Liouville distribution function [7] can then be defined as $D(\{X_i\};t)$ in the T-space as

$$D(\{X_i\};t)d\{X_i\} \equiv \lim_{n\to\infty}\frac{[No.of\ system\ points\ in\ the\ infinitesimal\ volume\ d\{X_i\}in\ T-space\ withing\ \{X_i\}]}{N} \tag{12}$$

which by definition satisfies the normalization condition.

http://dx.doi.org/10.5772/intechopen.70843

$$\int D(\{X_i\}; t) d\{X_i\} = 1. \tag{13}$$

The N system points distributed in the T-space do not interact with each other, behaving like an ideal gas. The distribution function $D(\{X_i\}; t)$ therefore satisfies a Liouville-type continuity equation

$$\frac{\partial D}{\partial t} + \{\dot{X}_i\}.\frac{\partial D}{\partial\{X_i\}} = 0 \tag{14}$$

The distribution is conserved along a trajectory in the phase-space distribution.

We can now perform a statistical averaging of a fine-grained quantity $A(X, X', \ldots\ldots; \{X_i(t)\}$ defined at a set of points $(X, X', \ldots)$ in the six-dimensional phase space. With the aid of Liouville distribution, we follow this way:

$$< A(X, X', \ldots\ldots; \mathrm{t}) > = \int d\{X_i\} D\{X_i\}; t) A(X, X', \ldots\ldots; \{X_i\}) \tag{15}$$

With respect to the conservation property, this average can be transformed equivalently into an average over the initial distribution, such that

$$< A(X, X', \ldots\ldots; \mathrm{t}) > = \int d\{X_i(0)\} D\{X_i(0)\}; t) A(X, X', \ldots\ldots; \{X_i(0)\}) \tag{16}$$

where $\{X_i(0)\}; t)\}$ represents the coordinates of the system points in T-space at t under the condition that it is located at $\{X_i(0)\}$ when $t = 0$.

2.3. BBGKY hierarchy

The distribution functions can be obtained through a statistical average of products of Klimontovich functions. A shorthand notation and numerals 1, 2, 3……, etc., in place of X, X', X''……., etc., can be used to simplify the presentations.

The Klimontovich Eq. (9) can therefore be written as

$$\left[\frac{\partial}{\partial t} + L(1)\right] N(1; t) = \int V(1, 2) N(1; t) N(2; t) d2. \tag{17}$$

The Liouville average of this equation can be obtained by using the methods in Eq. (16). The averaging process commutes with differential operators involved in Eq. (17). Now, with the aid of a single-particle distribution function

$$< N(X; \mathrm{t}) > = f_1(X; t) \tag{18}$$

The average of the second term defines the two-particle distribution function:

$$f_2(X,X';t);\ < N(X;t)N(X';t) >= \frac{1}{n}\delta(X - X')f_1(X;t) + f_2(X,X';t). \tag{19}$$

$$\left[\frac{\partial}{\partial t} + L(1)\right]f_1(1;t) = \int V(1,2)\left\{\frac{1}{n}\delta(1-2)f_1(1;t) + f_2(1,2;t)d2\right. \tag{20}$$

For an arbitrary function $y(1,2,\ldots;t)$, it can be proved from symmetry considerations that

$$\int V(1,2)\delta(1-2)y(1,2,..;t)d2 = 0 \tag{21}$$

Consequently,

$$\left[\frac{\partial}{\partial t} + L(1)\right]f_1(1;t) = \int V(1,2)f_2(1,2;t)d2 \tag{22}$$

It can start from an equation as well

$$\left[\frac{\partial}{\partial t} + L(1) + L(2)\right]N(1;t)N(2;t) = \int [V(1,3) + V(2,3)]N(1;t)N(2;t)N(3;t)d3 \tag{23}$$

Eq. (23) can be derived from a combination of Klimontovich equations. After averaging this equation with respect to the Liouville distribution and Eqs. (16) and (17), an equation involving f_1, f_2, and f_2 can be obtained. This equation can then be simplified with the aid of Eqs. (21) and (22), the result yields

$$\left\{\frac{\partial}{\partial t} + L(1) + L(2) - \frac{1}{n}[V(1,2) + V(2,1)]\right\}f_2(1,2;t) = \int [V(1,3) + V(2,3)]f_3(1,2,3;t)d3. \tag{24}$$

Similarly, it is considered that a Klimontovich equation for a product of an arbitrary number of the Klimontovich functions performs a statistical average of the equation. We therefore obtain the BBGKY hierarchy equations expressed as

$$\left[\sum_{i=1}^{s} L(i) - \frac{1}{n}\sum_{i\neq j}^{s} V(i,j)\right]f_s(1,\ldots,s;t)] = \sum_{i=1}^{s}\int \left[V(i,s+1)f_{s+1}(1,\ldots s+1;t)\right] d(s+1). \tag{25}$$

The set of equations in Eq. (25) is the basis for the kinetic theory of plasmas.

2.4. Vlasov's equation

For identical non-interacting particles, Liouville's equation can be written in T-space. Introducing two properties of identical non-interacting particles such as the distribution function f and the Hamiltonian function q simplifies the problem. The distribution function written as a function of $6N$ variables and time factorizes to a product of N functions, each involving only the coordinates and momenta of one particle, and time.

http://dx.doi.org/10.5772/intechopen.70843

The Hamiltonian function of N non-interacting particles is the sum of N terms, each involving only the coordinates and momenta of one particle. For identical particles, the terms of the Hamiltonian are also identical. For weakly or occasionally interacting particles, the decomposition of f into a product of factors and of q into a sum of terms is identical.

For a collection of N identical interacting particles, Liouville's theorem can be written in the μ-space as

$$\frac{\partial f}{\partial t}+\sum_{1}^{3}\left[\left(\frac{\partial f}{\partial x_j}\right)\left(\frac{\partial x_j}{\partial t}\right)+\left(\frac{\partial f}{\partial p_j}\right)\left(\frac{\partial p_j}{\partial t}\right)\right]=\frac{\partial f}{\partial t}=\left(\frac{\partial f}{\partial t}\right)_{int} \tag{26}$$

The right-hand side of Eq. (26) can be evaluated using the exact form of the interaction terms in the $6N+1$ Hamiltonian function variable. It is assumed that the Hamiltonian expression involves no magnetic terms. Under such conditions, $p=Mv$. If the coordinate system in the μ-space is changed from $x, y, z,$ p_x, p_y, p_z to $x, y, z,$ v_x, v_y, v_z, then the corresponding volume elements will be in the ratio M^3. Hence, if the figurative points density in the (r, p) space is constant according to Eq. (26),

$$\frac{\partial f}{\partial t}+\sum_{1}^{3}\left[\left(\frac{\partial f}{\partial x_j}\right)\left(\frac{\partial x_j}{\partial t}\right)+\left(\frac{\partial f}{\partial v_j}\right)\left(\frac{\partial v_j}{\partial t}\right)\right]=\frac{\partial f}{\partial t}=0 \tag{27}$$

If magnetic terms are included in the Hamiltonian function, then $p=Mv+qA$. When the coordinates are changed, the ratio of the corresponding volume elements can be calculated by using Jacobian. The Jacobian is a determinant calculated by taking the partial derivative of any coordinate of one system with respect to all the coordinates of the second system. For instance, in the physical space, the Jacobian is

$$\textbf{Jacobian}=\begin{vmatrix}\frac{\partial x_1}{\partial x_2} & \frac{\partial x_1}{\partial y_2} & \frac{\partial x_1}{\partial z_2}\\ \frac{\partial y_1}{\partial x_2} & \frac{\partial y_1}{\partial y_2} & \frac{\partial y_1}{\partial z_2}\\ \frac{\partial z_1}{\partial x_2} & \frac{\partial z_1}{\partial y_2} & \frac{\partial z_1}{\partial z_2}\end{vmatrix}$$

The value of the determinant is equal to the ratio of the corresponding volume elements. The ratio of the volume elements, M^3, is constant even in the presence of magnetic forces.

$\frac{\partial v_j}{\partial t}=a_j=F_j/\boldsymbol{M}$, where $\boldsymbol{a}$ = acceleration and $\boldsymbol{F}$ = external force. Introducing an operator ∇_v,

$$\nabla_v=\boldsymbol{i}\frac{\partial}{\partial v_x}+\boldsymbol{j}\frac{\partial}{\partial v_y}+\boldsymbol{k}\frac{\partial}{\partial v_z} \tag{28}$$

where $\boldsymbol{i}, \boldsymbol{j}$, and $\boldsymbol{k}$ are the unit vectors in the v_x, v_y, and v_z directions, respectively. In a vectorial form, Eq. (27) now becomes

$$\frac{\partial f}{\partial t} + v.\nabla f + \frac{\boldsymbol{F}}{\boldsymbol{M}}.\nabla_{\mathbf{v}} f = 0 \tag{29}$$

Eq. (29) is known as *Vlasov's equation*. *F* is the sum of the electric, magnetic, and gravitational forces resulting from external fields and the macroscopic forces resulting from the plasma itself. If we consider the viscous-like forces, Vlasov's equation becomes

$$\frac{\partial f}{\partial t} + v.\nabla f + \nabla_{\mathbf{v}}.\left[\frac{\boldsymbol{F}}{\boldsymbol{M}} f\right] = 0 \tag{30}$$

2.5. Maxwell's equations

Maxwell's equations express the relations between electric and magnetic fields in a medium. Consider a current ***I*** flowing in an element of length ***dl***. The magnetic flux density ***dB*** produced by this current at a point **P**, a distance ***r*** from the element, and making an angle θ with its axis is known as Ampere's law

$$\boldsymbol{dB} = KI\boldsymbol{dl} \sin\theta / r^2 \tag{31}$$

where *K* is a constant of proportionality defined as

$$K = \mu / 4\pi \tag{32}$$

where μ = permeability of the medium. For vacuum, $\mu = \mu_0 = 4\pi \times 10^{-7}$ H/m. The total magnetic flux density ***B*** produced at point **P** by the current flowing in a long conductor is

$$\boldsymbol{B} = \frac{\boldsymbol{\mu I}}{4\pi} \int \frac{\boldsymbol{\sin\theta}}{\boldsymbol{r^2}} dl \tag{33}$$

For infinite and linear conductor, $\boldsymbol{B} = \frac{\mu I}{2\pi r}$; where r = radial distance from **P** to the linear conductor as shown in **Figure 1**.

Figure 1. Flux density near a straight wire in which current *I* flows where $\sin\theta \boldsymbol{dl} = r d\theta$ and $r_0 = r \sin\theta$.

If ***B*** is integrated around the path that encloses the wire, then

$$\int \boldsymbol{B}.dl = \mu I \tag{34}$$

A magnetic field vector ***H*** is introduced to make the equation independent of the medium. It is such that

$$\boldsymbol{B} = \mu \boldsymbol{H} \tag{35}$$

Introducing Eq. (35) into Eq. (34) and the current ***I*** by the surface integral of the conduction current density ***J*** over the area described by the path of integration of ***H***,

http://dx.doi.org/10.5772/intechopen.70843

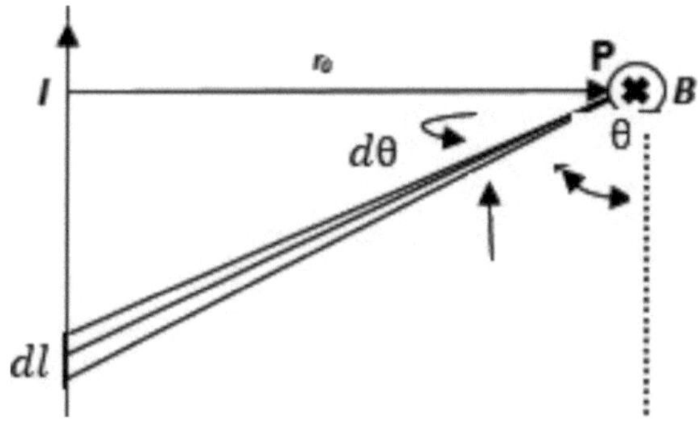

Figure 1. Flux density near a straight wire in which current I flows where $\sin\theta d\boldsymbol{l} = rd\theta$ and $r_0 = r\sin\theta$.

$$\int \boldsymbol{H}.dl = \iint_S \boldsymbol{J}.d\boldsymbol{S} \tag{36}$$

The total current density = conduction current density $\boldsymbol{\sigma E}$; $\boldsymbol{\sigma}$ = conductivity of the wire and $\boldsymbol{E}$ = electric field. $\boldsymbol{\partial D/\partial t}$ = displacement current density; $\boldsymbol{D}$ = electric flux density. Applying Stokes' theorem, Eq. (36) can now be written in a general form as

$$\nabla \times \boldsymbol{H} = \boldsymbol{J} + \frac{\partial \boldsymbol{D}}{\partial \boldsymbol{t}} \tag{37}$$

According to Faraday's law, the total e.m.f (V) induced in a closed loop as a function of the total flux $\boldsymbol{\Phi}_{\mathrm{m}}/\mathrm{d}\boldsymbol{t}$ producing the e.m.f is given as.

$$\mathrm{V} = -\,\mathrm{d}\boldsymbol{\Phi}_{\mathrm{m}}/\mathrm{dt} \tag{38}$$

If the total flux linkage $\boldsymbol{\Phi}_{\mathrm{t}}$ is $\mathrm{n}\boldsymbol{\Phi}_{\mathrm{m}}$, then $\mathrm{V} = -\mathrm{d}\boldsymbol{\Phi}_{\mathrm{t}}/\mathrm{d}\boldsymbol{t}$. The total flux through the circuit is equal to the integral of $\boldsymbol{B}$ over the area bounded by the circuit. Therefore,

$$\mathrm{V} = -\frac{\mathrm{d}}{\mathrm{d}t}\iint_S \boldsymbol{B}.d\boldsymbol{S} \tag{39}$$

The change in magnetic field produces an electric field $\boldsymbol{E}$, thus

$$V = \int \boldsymbol{E}.dl \text{ or } \boldsymbol{E} = -\nabla \boldsymbol{V} \tag{40}$$

Combining Eqs. (39) and (40), the induced e.m.f is

$$\int \boldsymbol{E}.dl = -\iint \left(\frac{\partial \boldsymbol{B}}{\partial t}\right).d\boldsymbol{S} \tag{41}$$

Applying Stokes' theorem on Eq. (41),

$$\nabla \times E = -\left(\frac{\partial B}{\partial t}\right) \tag{42}$$

Gauss' law states that the surface integral of the normal component of the electric flux density D over any closed surface equals the total enclosed charge q [8]. D is proportional to the electric field with permittivity ε of the medium as the constant of proportionality ($D \propto \varepsilon E$). $\varepsilon = \varepsilon_0$ in free space.

Replacing q with the integral of the charge density ϱ_e over the volume enclosed by the surface **S**, the vectorial form

$$\nabla . D = \varrho_e \tag{43}$$

For magnetic field, the integral of **B** over a closed surface is always equal to zero, thus

$$\nabla . B = 0 \tag{44}$$

Eqs. (37) and (42)–(44) are known as Maxwell's equations. The equations are satisfied in all plasma physics phenomena.

2.6. Liouville's theorem

In relation with the Boltzmann approach, most of the problems of statistical mechanics are best studied in multidimensional spaces called "phase spaces." Consider a μ-space, which is a six-dimensional and makes use of coordinates and the three components of momentum x, y, z, p_x, p_y, p_z or any set of Lagrangian coordinates for a point together with the associated generalized momenta. In this space, each plasma particle is represented by a point. If only one degree of freedom exists, the μ-space can be represented on a plane of **Figure 2**.

Figure 2. Trajectory of an oscillating point in the μ-space.

In the μ-space, the distribution function f is a function of seven variables. Thus, the probability of finding a particle in a given volume element depends only on the coordinates and momenta of this particle, not on those of the other particles. This is a simplified version of Liouville's theorem for a large number of non-interacting particles.

In the six-dimensional μ-space, the particles are conservative just as they are in the ordinary space. Hence, the conservation theorem may be applied in the phase space,

$$\frac{\partial B}{\partial t} + \nabla_6 . (f v_6) = 0 \tag{45}$$

where ∇_6 = six-dimensional divergence and v_6 is a six-dimensional velocity vector whose components are exact time derivatives of the six coordinates of the μ-space. We now write ∇ temporarily x_1, x_2, x_3 instead of x, y, z,

http://dx.doi.org/10.5772/intechopen.70843

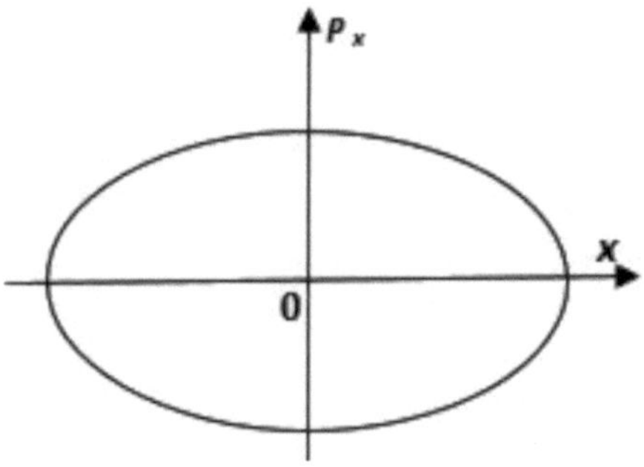

Figure 2. Trajectory of an oscillating point in the μ-space.

$$\frac{\partial B}{\partial t} + \sum_{1}^{3} \left[\frac{\partial (f \dot{x}_j)}{\partial x_j} + \frac{\partial \left(f \dot{p}_j \right)}{\partial p_j} \right] = 0 \tag{46}$$

or

$$\frac{\partial f}{\partial t} + \sum_{1}^{3} \left[\frac{\partial f}{\partial x_j} \dot{x}_j + \frac{\partial f}{\partial p_j} \dot{p}_j + f \left(\frac{\partial \dot{x}_j}{\partial x_j} + \frac{\partial \dot{p}_j}{\partial p_j} \right) \right] = 0 \tag{47}$$

Hamilton's canonical equations yield

$$\frac{\partial q_j}{\partial q_j} + \frac{\partial \dot{p}_j}{\partial p_j} = \frac{\partial}{\partial q_j} \left(\frac{\partial q}{\partial p_j} \right) - \frac{\partial}{\partial p_j} \left(\frac{\partial q}{\partial q_j} \right) = \frac{\partial^2 q}{\partial p_j \partial q_j} - \frac{\partial^2 q}{\partial q_j \partial p_j} = 0 \tag{48}$$

where q = Hamiltonian. Eq. (47) now becomes

$$\frac{\partial f}{\partial t} + \sum_{1}^{3} \left[\left(\frac{\partial f}{\partial x_j} \right) \left(\frac{\partial x_j}{\partial t} \right) + \left(\frac{\partial f}{\partial p_j} \right) \left(\frac{\partial p_j}{\partial t} \right) \right] = \frac{\partial f}{\partial t} = 0 \tag{49}$$

Eq. (49) is known as *Liouville's equation.*

2.7. Boltzmann's equation

The Boltzmann equation provides the statistical analysis of all the individual positions and momenta of each particle in the fluid (macro-system) at an instant, that is, number of particles in a particular level and their distribution among different levels [9]. It gives relative number of atoms in different excitation states as a function of temperature and refers to certain number of atoms or ions in a particular excitation state with respect to the ground state. Boltzmann equation gives a mathematical description of the state of a system and how it changes. It describes a quantity called the distribution function, f, which depends on a position, velocity, and the time. The function f determines the average number of particles having velocities within a small range from v to $v + \Delta v$ and coordinates within a small range from r to $r + \Delta r$ in time Δt.

In a hot dense gas, the atoms constantly experience collisions with each other, which lead to excitation to the different possible energy levels. The collisional excitation follows radiative de-excitation in timescales of the order of nanoseconds. For a constant temperature and pressure, a dynamic equilibrium is established between collisional excitations and radiative de-excitations, which lead to particular distribution of the atoms among different energy levels. Most of the atoms are at low-lying levels. The number of atoms at higher levels decreases exponentially with energy level. At low temperature, the faster the population drops at the higher levels. Only at very high temperatures, high-lying energy levels are occupied by an appreciable number of atoms. Boltzmann's equation gives the distribution of the atoms among the various energy levels as a function of energy and temperature.

Let us consider a system at local thermal equilibrium (LTE) with a constant volume consisting of "N" atoms and each of atoms has "m" possible energy levels. Suppose there are "N_j" atoms in energy level "E_j." The total number N of atoms is given as

$$N = \sum_{i=1}^{m} N_i \tag{50}$$

The total energy "E" of the system can be written as

$$E = \sum_{i=1}^{m} N_i E_i \tag{51}$$

A number of ways in which "N_1" atoms from total atoms "N" can occupy the first level are $\binom{N}{N_1}$. In the same manner, the total number of ways to arrange "N_2" atoms from the remaining "N-1" atoms is $\binom{N-1}{N_2}$ and so on. Thus, the total number of microstates "X" in the system, that is, the number of ways to arrange "N" atoms of the system, is given as

$$X = \binom{N}{N_1}\binom{N-1}{N_2}\binom{N-2}{N_3}\ldots\ldots\ldots\binom{N-(m-1)}{N_m} \tag{52}$$

By solving the above binomial

$$X = \frac{N!}{N_1!N_2!\ldots\ldots N_j!\ldots\ldots N_m!} \tag{53}$$

$$X = \frac{N!}{\prod_{i=1}^{m} N_i!} \tag{54}$$

Taking log on both sides of Eq. (53),

http://dx.doi.org/10.5772/intechopen.70843

$$\ln X = \ln N! - \ln N_1! - lnN_2!\ldots\ldots - lnN_j!\ldots\ldots\ldots - lnN_m! \tag{55}$$

By applying Stirling's approximations to the factorials of all variables,

$$\ln X \cong \ln N! - (N_1 \ln N_1 - N_1) - (N_2 lnN_2 - N_2)\ldots\ldots \tag{56}$$

$$\ln X \cong \ln N! - \sum_{i=1}^{m} N_i \ln N_i + N \tag{57}$$

Let us maximize the ln$X$ with respect to one microstate "N_j," in a manner that is consistent with constrains of Eqs. (50) and (51). Lagrangian multiplier for the most probable occupation of the jth level is given as

$$\frac{\partial lnX}{\partial N_j} + \lambda \frac{\partial N}{\partial N_j} + \mu \frac{\partial E}{\partial N_j} = 0 \tag{58}$$

where μ and λ are Lagrangian's multipliers.

By adding values from Eqs. (50), (51), and (56) in Eq. (58),

$$\frac{\partial [lnN! - (N_1 \ln N_1 - N_1) - (N_2 lnN_2 - N_2)\ldots\ldots]}{\partial N_j} + \lambda \frac{\partial (N_1 + \ldots + N_j \ldots)}{\partial N_j} + \mu \frac{\partial (E_1 N_1 + \ldots + E_j N_j \ldots)}{\partial N_j} = 0 \tag{59}$$

On operating differential,

$$-\ln N_j + \lambda + \mu E_j = 0 \tag{60}$$

$$\lambda + \mu E_j = \ln N_j \tag{61}$$

$$N_j = e^{\lambda + \mu E_j} \tag{62}$$

$$N_j = e^{\lambda} e^{\mu E_j} \tag{63}$$

$$N_j = C e^{\mu E_j} \tag{64}$$

In general form by multiplying both sides by "N_j", Eq. (60) can be written as

$$-\sum_{i=1}^{m} N_i \ln N_i + N\lambda + \mu E = 0 \tag{65}$$

From Eqs. (57) and (65),

$$\ln X - \ln N! - N + N\lambda + \mu E = 0 \tag{66}$$

$$\ln X = \ln N! - (\lambda + 1)N - \mu E \tag{67}$$

The change in the internal energy of the system in terms of thermodynamics equation can be written as

$$dE = TdS - PdV \tag{68}$$

where "T" is the temperature, "P" is the pressure, and "S" is the entropy of the system.

The change in the internal energy at a constant volume is given as

$$\left(\frac{\partial E}{\partial S}\right)_V = T \tag{69}$$

$$\left(\frac{\partial S}{\partial E}\right)_V = \frac{1}{T} \tag{70}$$

Boltzmann's equation for entropy is

$$S = k \ln X \tag{71}$$

$$S = k \ln X \tag{72}$$

From Eqs. (67) and (72),

$$S = k\left[\ln N! - (\lambda + 1)N - \mu E\right] \tag{73}$$

Differentiate with respect to the total energy at a constant volume

$$\left(\frac{\partial S}{\partial E}\right)_V = k[-\mu] \tag{74}$$

By comparing Eqs. (70) and (74)

$$\frac{1}{T} = k[-\mu] \tag{75}$$

$$\mu = -\frac{1}{kT} \tag{76}$$

By adding value of "μ" in Eq. (64)

$$N_j = Ce^{-\frac{E_j}{kT}} \tag{77}$$

To calculate the value of "C", let us change the subscript "j" to "i" for Eq. (77) and take summation from "1" to "m".

$$N = C\sum_{i=1}^{m} e^{-\frac{E_i}{kT}} \tag{78}$$

http://dx.doi.org/10.5772/intechopen.70843

and

$$C = \frac{N}{\sum_{i=1}^{m} e^{-\frac{E_i}{KT}}} \tag{79}$$

Thus, Eq. (64) can be written as

$$N_j = \frac{N e^{-\frac{E_j}{KT}}}{\sum_{i=1}^{m} e^{-\frac{E_i}{KT}}} \tag{80}$$

$$\frac{N_j}{N} = \frac{e^{-\frac{E_j}{KT}}}{\sum_{i=1}^{m} e^{-\frac{E_i}{KT}}} \tag{81}$$

In a system, most of the energy levels in an atom are degenerated, that is, atoms have several states with the same energy. To find out the population of an atom at a particular level, the population of each constituent state is required to be added together. Thus, each term in Eq. (81) must be multiplied by the statistical weight (degeneracy) "ϖ" of the level

$$\frac{N_j}{N} = \frac{\varpi_j e^{-\frac{E_j}{KT}}}{\sum_{i=1}^{m} \varpi_i e^{-\frac{E_i}{KT}}} \tag{82}$$

The term "$\sum_{i=1}^{m} \varpi_i e^{-\frac{E_i}{KT}}$" is called the partition function. The Eq. (82) gives the relative number of atoms in state "j" with respect to the total number of atoms in the system. The number of atoms in level "j" can also be compared with the number of atoms at the ground level

$$\frac{N_j}{N_o} = \frac{\varpi_j e^{-\frac{E_j}{KT}}}{\varpi_o e^{-\frac{E_o}{KT}}} \tag{83}$$

$$N_j = \frac{N_o \varpi_j e^{-\frac{E_j}{KT}}}{U} \tag{84}$$

and the number of atoms in level 2 relative to level 1, where level 2 is higher than level 1

$$\frac{N_2}{N_1} = \frac{\varpi_2 e^{-\frac{E_2}{KT}}}{\varpi_1 e^{-\frac{E_1}{KT}}} \tag{85}$$

$$\frac{N_2}{N_1} = \frac{\varpi_2}{\varpi_1} e^{-\frac{E_2 - E_1}{KT}} \tag{86}$$

The Einstein A coefficient gives the probability of spontaneous emission. A quantum of radiation is emitted by an atom when it de-excites from an excited level to a lower level, which is given as

$$h\nu = \Delta E \tag{87}$$

where "ν" is the frequency of emitted radiation and "$\Delta E = E_2$-E_1" is the energy difference between two atomic states (or level) "E_2" higher level and "E_1" lower level.

Suppose the number of downward transitions per unit time is merely proportional to the number of atoms "N_2" at a higher state, then, the number of transition per unit time is given as

$$-\dot{N_2} = A_{21}N_2 \tag{88}$$

where "A_{21}" is proportionality constant known as Einstein coefficient for spontaneous emission for transition from level "E_2" to level "E_1".

As "$A_{21}N_2$" is the downward transition per unit time from "E_2" to "E_1", thus the rate of emission of energy from these "N_2", that is, radiant power or flux, is given as

$$\Phi = A_{21}N_2 h\upsilon_2 \tag{89}$$

Here, "ν_{21}" represents the frequency of radiation due to transition from level "E_2" to level "E_1". As the radiation is emitted isotropically, thus the intensity is

$$\mathrm{I}_{21} = \frac{A_{21}N_2 h\upsilon_{21}}{4\pi} \tag{90}$$

From Eqs. (84) and (90)

$$\mathrm{I}_{21} = \frac{N_o A_{21}\varpi_2 h\upsilon_{21}}{4\pi U} e^{-\frac{E_2}{kT}} \tag{91}$$

$$\mathrm{I}_{21} = \frac{N_o A_{21}\varpi_2 hc}{4\pi\lambda_{21} U} e^{-\frac{E_2}{kT}} \tag{92}$$

2.7.1. Boltzmann plot

Taking log and solving Eq. (92),

$$\ln\left(\frac{\mathrm{I}_{21}\lambda_{21}}{A_{21}\varpi_2}\right) = -\frac{E_2}{kT} + \ln\left(\frac{N_o hc}{4\pi U}\right) \tag{93}$$

2.7.2. Intensity ratio method

Consider two different emission lines from level $i \to j$ and $m \to n$, where i and m are higher energy levels, j and n are lower energy levels. By using Eq. (92),

$$\mathrm{I}_{ij} = \frac{N_o A_{ij}\varpi_i hc}{4\pi\lambda_{ij} U} e^{-\frac{E_i}{kT}} \tag{94}$$

http://dx.doi.org/10.5772/intechopen.70843

$$\mathrm{I}_{mn} = \frac{N_o A_{mn} \varpi_m hc}{4\pi \lambda_{mn} U} e^{-\frac{E_m}{kT}} \tag{95}$$

Taking ratio of Eqs. (94) and (95) and solving for temperature "T",

$$\ln\left(\frac{\mathrm{I}_{ij}\lambda_{ij}A_{mn}\varpi_m}{\mathrm{I}_{mn}\lambda_{mn}A_{ij}\varpi_i}\right) = -\frac{E_i - E_m}{kT} \tag{96}$$

$$\mathrm{T} = -\frac{E_i - E_m}{kln\left(\frac{I_{ij}\lambda_{ij}A_{mn}\varpi_m}{I_{mn}\lambda_{mn}A_{ij}\varpi_i}\right)} \tag{97}$$

2.8. Saha equation

The Boltzmann equation gives only the relative number of atoms or ions in a particular excitation state with respect to the ground state and it does not provide the total number of atoms that have been ionized. In order to determine the total abundance of a given element, it is necessary to know how the atoms are distributed among their several ionization stages. To quantify the number of atoms/ions in different ionization states, Saha's equation is used which gives an expression for the total number of ions in an ionization state relative to lower ionization state.

For a system at local thermal equilibrium with a constant volume, the Boltzmann equation for the number of ions "N^i" relative to atoms/ions "N^i_1" in the ground state of ionization state "i" can be written as (by using Eq. (85))

$$\frac{N^i}{N^i_1} = \frac{\varpi^i e^{-\frac{E^i}{kT}}}{\varpi^i_1 e^{-\frac{E^i_1}{kT}}} \tag{98}$$

Taking the sum of all the excited states "j" of ionization state "i", Eq. (98) can be written as

$$\frac{N^i_j}{N^i_1} = \frac{\sum_{j=1}^{\infty} \varpi^i_j e^{-\frac{E^i_j}{kT}}}{\varpi^i_1 e^{-\frac{E^i_1}{kT}}} \tag{99}$$

$$\frac{N^i_j}{N^i_1} = \frac{\sum_{j=1}^{\infty} \varpi^i_j e^{-\frac{E^i_j}{kT}}}{Z_1} \tag{100}$$

where $Z_1 = \varpi^i_1 e^{-\frac{E^i_1}{kT}}$, which represents the number of atoms at the ground level. The same expression for the ionization state "$i + 1$" will include not only the excitation states of ion "$i + 1$" but also the free electrons. Let us consider the energy of the free electron "E_e" and momentum "P_e" with "$E_e = P_e^2/2m_e$". Then, at any state of the system of ion "$i + 1$" and an electron is characterized by the total energy "$E^{i+1}_j - E_e$" and statistical weight "$\varpi^{i+1}_j \varpi_e(P_e)$". As the energies of free electrons are continuous, the Boltzmann expression analogous to Eq. (100) over all possible momenta of free electron can be written as

$$\frac{N_j^{i+1}}{N_1^i} = \frac{1}{\varpi_1^i} \sum_{j=1}^{\infty} \varpi_j^{i+1} e^{-\frac{\left(E_j^{i+1} - E_1^i\right)}{kT}} \int_0^{\infty} \varpi_e(P_e) e^{-\frac{E_e}{kT}} dP_e \tag{101}$$

The ionization potential of the ion "i" can be defined as

$$\chi_i = E_1^{i+1} - E_1^i \tag{102}$$

Thus, we can write

$$E_j^{i+1} - E_1^i = E_j^{i+1} - E_1^{i+1} + \chi_i \tag{103}$$

Eq. (101) will become

$$\frac{N_j^{i+1}}{N_1^i} = \frac{1}{\varpi_1^i} \sum_{j=1}^{\infty} \varpi_j^{i+1} e^{-\frac{\left(E_j^{i+1} - E_1^{i+1} + \chi_i\right)}{kT}} \int_0^{\infty} \varpi_e(P_e) e^{-\frac{E_e}{kT}} dP_e \tag{104}$$

$$\frac{N_j^{i+1}}{N_1^i} = \frac{1}{\varpi_1^i e^{-\frac{E_1^{i+1}}{kT}}} \sum_{j=1}^{\infty} \varpi_j^{i+1} e^{-\frac{\left(E_j^{i+1} + \chi_i\right)}{kT}} \int_0^{\infty} \varpi_e(P_e) e^{-\frac{E_e}{kT}} dP_e \tag{105}$$

$$\frac{N_j^{i+1}}{N_1^i} = \frac{1}{Z_1} \sum_{j=1}^{\infty} \varpi_j^{i+1} e^{-\frac{\left(E_j^{i+1} + \chi_i\right)}{kT}} \int_0^{\infty} \varpi_e(P_e) e^{-\frac{E_e}{kT}} dP_e \tag{106}$$

where $Z_1 = \varpi_1^i e^{-\frac{E_1^{i+1}}{kT}}$, which represents the number of atoms at the ground level. Although the momenta of the free electrons have a continuous distribution, but according to Heisenberg's uncertainty principle the electrons within a phase-space volume ΔV are indistinguishable, unless they have an opposite spin orientation, that is,

$$\Delta V\, P_e^3 = h^3 \tag{107}$$

There are two possible distinguishable electron states within phase space of $\Delta V\, P_e^3 = h^3$. Thus, the statistical weight for the free electron can be written as

$$\varpi_e(P_e) dP_e = 2 \frac{\Delta V\, P_e^3}{h^3} \tag{108}$$

If the distribution of electron momenta is isotropic, then

$$\Delta P_e^3 = 4\pi P_e^2 dP_e \tag{109}$$

http://dx.doi.org/10.5772/intechopen.70843

If "n_e" is the number density of electron, then the differential volume per electron will be

$$\Delta V = \frac{1}{n_e} \tag{110}$$

Thus, Eq. (108) can be written as

$$\varpi_e(P_e)dP_e = \frac{8\pi P_e^2}{n_e h^3} dP_e \tag{111}$$

We can write the integral in Eq. (106) as

$$\int_0^\infty \varpi_e(P_e)e^{-\frac{E_e}{kT}}dP_e = \frac{8\pi}{n_e h^3}\int_0^\infty P_e^2 e^{-\frac{E_e}{kT}}dP_e \tag{112}$$

The integral can be transformed into an integral over variable "z" such that

$$z = \frac{E_e}{kT} = \frac{P_e^2}{2m_e KT} \tag{113}$$

$$P_e = \sqrt{2m_e KTZ} \tag{114}$$

$$dP_e = \frac{1}{2}\sqrt{\frac{2m_e KT}{z}}dz \tag{115}$$

By replacing values of P_e and dP_e in Eq. (112)

$$\int_0^\infty \varpi_e(P_e)e^{-\frac{E_e}{kT}}dP_e = \frac{4\pi}{n_e h^3}(2m_e KT)^{3/2}\int_0^\infty \sqrt{z}e^{-z}dz \tag{116}$$

The integral on the right-hand side of Eq. (116) is the Gamma function of the argument 3/2, which is $\Gamma(3/2) = (1/2)\Gamma(1/2) = \sqrt{\pi}/2$. By adding this value in Eq. (116)

$$\int_0^\infty \varpi_e(P_e)e^{-\frac{E_e}{kT}}dP_e = \frac{4\pi}{n_e h^3}(2m_e KT)^{3/2}\left(\sqrt{\pi}/2\right) \tag{117}$$

$$\int_0^\infty \varpi_e(P_e)e^{-\frac{E_e}{kT}}dP_e = \frac{2}{n_e h^3}(2\pi m_e KT)^{3/2} \tag{118}$$

From Eqs. (106) and (118),

$$\frac{N_j^{i+1}}{N_1^i} = \frac{1}{Z_1}\sum_{j=1}^{\infty}\varpi_j^{i+1}e^{-\frac{\left(E_j^{i+1}+\chi_i\right)}{kT}}\frac{2}{n_e h^3}(2\pi m_e KT)^{3/2} \tag{119}$$

By dividing Eqs. (100) and (119),

$$\frac{N_j^{i+1}}{N_j^i} = \frac{\sum_{j=1}^{\infty} \varpi_j^{i+1} e^{-\frac{\left(E_j^{i+1}+\chi_i\right)}{kT}}}{\sum_{j=1}^{\infty} \varpi_j^i e^{-\frac{E_j^i}{kT}}} \frac{2}{n_e h^3} (2\pi m_e KT)^{3/2} \tag{120}$$

$$\frac{N_j^{i+1}}{N_j^i} = \frac{\sum_{j=1}^{\infty} \varpi_j^{i+1}}{\sum_{j=1}^{\infty} \varpi_j^i} \frac{2}{n_e h^3} (2\pi m_e KT)^{3/2} e^{-\frac{\left(E_j^{i+1}+\chi_i-E_j^i\right)}{kT}} \tag{121}$$

$$\frac{N_j^{i+1}}{N_j^i} = \frac{\sum_{j=1}^{\infty} \varpi_j^{i+1}}{\sum_{j=1}^{\infty} \varpi_j^i} \frac{2}{n_e h^3} (2\pi m_e KT)^{3/2} e^{-\frac{\left(E_j^1-E_j^i+\chi_i\right)}{kT}} \tag{122}$$

For a particular state,

$$\frac{N^{i+1}}{N^i} = \frac{2(2\pi m_e KT)^{3/2}}{n_e h^3} \frac{\varpi^{i+1}}{\varpi^i} e^{-\frac{\left(E^{i+1}-E^i+\chi_i\right)}{kT}} \tag{123}$$

$$n_e = \frac{2(2\pi m_e KT)^{3/2}}{h^3} \frac{N^i \varpi^{i+1}}{N^{i+1} \varpi^i} e^{-\frac{\left(E^{i+1}-E^i+\chi_i\right)}{kT}} \tag{124}$$

Eq. (90) can be written as

$$N^{i+1} = \frac{4\pi h I^{i+1} \lambda^{i+1}}{A^{i+1} c} \tag{125}$$

$$N^i = \frac{4\pi h I^i \lambda^i}{A^i c} \tag{126}$$

From Eqs. (124)–(126),

$$n_e = \frac{2(2\pi m_e KT)^{3/2}}{h^3} \frac{I^i \lambda^i A^{i+1} \varpi^{i+1}}{I^{i+1} \lambda^{i+1} A^i \varpi^i} e^{-\frac{\left(E^{i+1}-E^i+\chi_i\right)}{kT}} \tag{127}$$

3. Summary

Kinetic theory provides the essential material for an introductory course on plasma physics as well as the basis for advanced kinetic theory. It offers a wide-range coverage of the field. Plasma kinetics deals with the relationship between velocity and forces and the study of continua in velocity space. The understanding of the most important plasma parameters, that is, plasma oscillations, plasma frequency, Debye shielding, Debye length, plasma temperature, and electron density, is important for studying plasmas. Kinetic theory has a wide scope. Plasma kinetics mathematical equations that will aid the readers in understanding simple techniques on how to investigate plasma dynamics and kinetics are discussed in this chapter.

http://dx.doi.org/10.5772/intechopen.70843

Kinetic theory provides the basics and essential introduction to plasma physics and subsequently advanced kinetic theory. Plasma understanding, waves, oscillations, frequencies, and applications are covered in kinetic theory. In fact, we have attempted to present a precise discussion of plasma kinetic theory which includes the basic plasma parameters mathematical formulation for an easy understanding to the reader.

Author details

Kashif Chaudhary*, Auwal Mustapha Imam, Syed Zuhaib Haider Rizvi and Jalil Ali

*Address all correspondence to: kashif@utm.my

Laser Center, Ibnu Sina Institute for Scientific and Industrial Research (ISU-SIR), Universiti Teknologi Malaysia (UTM), Malaysia

References

[1] Bittencourt JA. Fundamentals of Plasma Physics. Munich, Germany: Springer Science & Business Media; 2013

[2] Sitenko AG. Fluctuations and Non-Linear Wave Interactions in Plasmas: International Series in Natural Philosophy. Brazil: Elsevier; 2016. Vol. 107

[3] Thompson WB. An Introduction to Plasma Physics. Brazil: Elsevier, 2013

[4] Akhiezer AI, Akhiezer I, Polovin R. Collective Oscillations in a Plasma: International Series of Monographs in Natural Philosophy. Brazil: Elsevier; 2017. Vol. 7

[5] Davidson R. Methods in Nonlinear Plasma Theory. Brazil: Elsevier; 2012

[6] Klimontovich YL. The Statistical Theory of Non-Equilibrium Processes in a Plasma: International Series of Monographs in Natural Philosophy. Brazil: Elsevier; 2013. Vol. 9

[7] Rice SA, Gray P. The Statistical Mechanics of Simple Liquids. Olney, Bucks, UK: Interscience Publ; 1965

[8] Halliday D, Resnick R, Walker J. Principles of Physics Extended, International Student Version. India: Wiley India Pvt. Ltd.; 2016

[9] Thompson CJ. Mathematical Statistical Mechanics. Princeton, New Jersey, USA: Princeton University Press; 2015